Aquaculture Engineering

Basic Manual

THE AUTHORS

Dr. Menaga Meenakshisundaram is presently working as Assistant Professor in the Department of Aquaculture under Tamil Nadu Fisheries University, India. Acquired Ph.D. in Aquaculture from the Tamil Nadu Fisheries University and Master of Fisheries Science in Aquaculture at Central Institute of Fisheries Education, Mumbai. Serving as Student Director of World Aquaculture Society-Asian Pacific Chapter from 2019-2022. Associated as an external expert in disseminating the basics of Aquaculture to the USAID funded project. Five years of experience in teaching and as a mentor guided the post graduate students for their research in Aquaculture Worked as a Technical Associate with the International and National companies in Aquaculture by conducting the field trials. Recipient of "Mgr. Dr. Jan D. F. Heine Memorial Award 2019" in recognition of outstanding efforts to stimulate sustained improvement of worldwide human living standards by International Tilapia Foundation. She has international exposure on Aquaculture practices in Indonesia, Taiwan, Malaysia, Myanmar and Singapore. Served as National Organizing Secretary for conducting International and National conferences, seminars and workshops. As a team member, formulated and prepared many project proposals in Aquaculture to the funding for the National government agencies. Authored more than fifteen research papers and co-authored more than twenty technical proceedings of the National and International events. Actively involved in undergoing research in the trust areas of aquaculture such as Aquaculture systems (Raceways, RAS, Aquaponics etc.,), Biofloc technology, functional feed ingredients and digital application in aquaculture.

Prof. S. Felix has 38 years of teaching and research experience in aquaculture. He is the Past President of World Aquaculture Society-Asian Pacific Chapter and he has also served as Director of the World Aquaculture Society(APC).Prof.S.Felix has served as Second Vice Chancellor of Tamil Nadu Dr. J. Jayalalithaa Fisheries University and also served as Dean of the Fisheries College in Tamilnadu, India. He has developed over a dozen off-campus facilities and 40 Constituent units for the University all over the state for the benefit of students & farmers, besides creating half a dozen new Fisheries Institutes to impart fisheries education. He developed and introduced the raceway culture technology for the first time in India. Under his effective leadership, he has established the Aquatic Rainbow Technology Park and an aquatic Rainbow Mall to the tune of Rs.12 crores for ornamental fish culture as first of its kind in the country. He has 10 patents published on various aquaculture technologies to his credit. He has published 15 Text Books, 50 Manuals and more than 150 Research papers and reports. He has organised 80 Seminars and Conferences . Introduced 30 new courses comprising UG, PG , Ph.D and Paraprofessional programs. Implemented 30 R&D and Technology projects to develop a dozen aquaculture technologies for the benefit of farmers.

Aquaculture Engineering
Basic Manual

– Authors –

Menaga Meenakshisundaram
Sugantham Felix

2023
Daya Publishing House®
A Division of
Astral International Pvt. Ltd.
New Delhi – 110 002

ISBN: 9789359192055 (HB)

Published by : **Daya Publishing House®**
A Division of
Astral International Pvt. Ltd.
– ISO 9001:2015 Certified Company –
4736/23, Ansari Road, Darya Ganj
New Delhi-110 002
Ph. 011-43549197, 8130496929
e-mail: info@astralint.com
Website: www.astralint.com

Preface

Aquaculture Engineering encompasses an extremely wide array of application in Fisheries sector. This branch of engineering has become the forefront of global innovation for increasing the fish and shrimp production. Aquaculture Engineering has been around for more than decade and is currently approached by many fisheries professionals.

Going up a notch from the previous publications by various aquaculture and engineering professionals this publication brings Modern Aquaculture Engineering concepts. This comprehensive book on basic Aquaculture engineering covers almost all the topics of basic aquaculture engineering, ranging from site selection to traditional culture systems to advanced culture systems. An introduction to basic aquaculture engineering has also been provided so as to prepare the students for an in-depth study later. Easy explanation of topics, plenty of examples and illustrations, practice problems and multiple-choice questions with answers, are the principal features of this book. It is an excellent reference for students with a background in design and construction of farm including aeration, filtration and pumps.

The book features a general layout of fifteen chapters, basic style of engineering drawings, and worked problems. Engineers will also appreciate the author's approach in linking actual industrial setups and the engineering principles behind them. The authors also took extra effort to present the content in a precise way for the easy learning of the students to understand the core concepts of aquaculture engineering.

Authors are very sure as this book can be used as a great starting point by anyone who is studying Aquaculture Engineering and can also be used as a ready reference for under graduate as well as post graduate studies.

Menaga Meenakshisundaram
Sugantham Felix

Contents

Chapter 1

Site Selection Criteria for Fish Farms

A. FISH FARMS

I. Definition

Fish farm is a set of scientifically planned, designed and constructed ponds for various fish culture activities like breeding, hatching, rearing, nursing and stocking of fishes.

II. Objectives of Fish Farms

- ☆ For breeding of fishes
- ☆ To maintain brooders
- ☆ To conduct the research to improve fish variety and their culture
- ☆ To make facilities to store required quantity of fish seeds
- ☆ To demonstrate the fish culture activities to fish culturist

The success of fish culture is mainly depending upon proper planning and construction of pond. The design of the fish ponds are very much influenced by the type of pond and the type of culture to be practiced in the pond.

III. Classification of Fish Farm

Fish farms can be classified into different types based on various factors:

a) Based on kind of Materials Used for Enclosure

- ☆ Plastic tanks
- ☆ Cement concrete tanks
- ☆ FRP (Fibre Reinforced Plastic) tanks
- ☆ Liner ponds
- ☆ Earthen ponds

b) Based on Water Source

- ☆ Rain fed farm
- ☆ Tide fed farm
- ☆ Sewage water
- ☆ Groundwater
- ☆ Spring water
- ☆ Municipal/corporation water
- ☆ River/canal/dam

c) Based on Culture

- ☆ Extensive
- ☆ Semi-intensive
- ☆ Intensive
- ☆ Super-intensive
- ☆ Traditional culture

d) Based on the Characteristics of the Farm Environment

- ☆ Warm water
- ☆ Cold water

e) Based on Water Salinity

- ☆ Freshwater
- ☆ Brackish water
- ☆ Marine water

Freshwater is defined when the salt content is less than 0.5 ppt, brackish water is defined when the salt content is in the range of 0.50 -30 ppt, marine water is defined when the salt content is more than 30-35 ppt.

f) Based on Water Replacement

- ☆ Running
- ☆ Stagnant

g) Based on Physiographical Zone

- ☆ Inland fish farm
- ☆ Coastal fish farm
- ☆ Marine fish farm

IV. Different Systems of Aquaculture

An aquaculture unit is to develop a habitat within a body of water where aquatic organisms can be reared under the best possible growth conditions. Various techniques of aquaculture have been developed all over the world for sustainable yield of aquatic organisms. The aquaculture production systems may be grouped on the basis of the habitat, economic considerations, kind of fish farm operations, sex and specially adaptive fishes, climatic conditions, *etc.*

a) Aquaculture Systems Based on Habitat

- ☆ Freshwater culture system
- ☆ Brackish water culture system
- ☆ Mariculture system

b) Aquaculture Systems Based on Economic Considerations

i) Extensive Culture

- ☆ Culture management is minimum or almost nil
- ☆ Very low stocking density
- ☆ Aquatic organisms are raised on natural food
- ☆ Yield is in lower side
- ☆ Investment is very less

ii) Semi-intensive Culture

- ☆ Culture management includes supplementary feed along with natural food and fertilizer management
- ☆ Stocking density is more than that of extensive system
- ☆ Aeration supplied as and when required
- ☆ Yield is more than that of extensive system
- ☆ Investment is relatively more

iii) Intensive Culture

- ☆ Culture management is for maximum production from minimum quantity of water
- ☆ Stocking density is very high

- ☆ Provision of supplementary feed, fertilizer and aeration through out the culture period
- ☆ Health management is taken care
- ☆ Yield is maximum
- ☆ Investment is quite high

c) Aquaculture Systems Based on kind of Fish Farm Operations

i) **Complete farming:** This includes breeding the aquatic organisms and raising the seed. It involves many ponds of various designs. Generally three types of ponds are used for seed raising.

ii) **Restricted farming:** It involves one pond operation for either seed raising or raising to table size fish.

d) Aquaculture Systems Based on Sex and Specially Adaptive Fishes

i) **Monoculture**: Culture of single species *e.g.,* Shrimp culture

ii) **Polyculture** of 2 or more species together. Here, the species of different feeding habits are selected and stocked. *e.g.,* Composite fish culture of Indian Major carp (IMC) *i.e.* Catla, Rohu and Mrigal

iii) **Monosex culture**: Culture of one sex either male or female of one species *e.g.,* Tilapia.

iv) **Integrated farming**: Culture of fish along with other animals and/or crops *e.g.,* Rice- fish farming, Fish-poultry farming and Fish-duck farming

f) Aquaculture Systems Based on Climatic Conditions

i) **Warm water culture:** Culture practices in the water bodies of tropical climate.

ii) **Cold water culture:** Culture practices in the water body of higher altitudes.

V. Selection of Site for Aquafarm

Some of the suitable site selection criteria for the freshwater fish farm includes:

- ☆ Water supply and quality of water
- ☆ Soil
- ☆ Topography
- ☆ Biological and operational factors
- ☆ Meteorological factors
- ☆ Social factors
- ☆ Economic factors
- ☆ Legal factors

A. WATER SUPPLY AND QUALITY OF WATER

Water Supply

Sufficient quantity of good quality water supply is the single most important factor to be considered for selecting a suitable site for an aquaculture farm. For this, while considering a site for fish farm, a thorough investigation of the water supply is preferred so that water will be available at all times.

Total Water Requirement for a Fish Pond

The annual requirement of water for the fish ponds will be decided by:

a. Species to be cultured
b. Type of culture techniques to be followed
c. Seepage factor
d. Rate of evaporation and
e. Meteorological factors

Total quantity of water required for a pond on an average per year is calculated using the following formula

$$QTw = (Vp + V rf + Le + Ls + Lc - Vra)\ m^3.$$

Source

Depending upon the nature of occurrence, water sources can be classified into following groups.

a. **Surface water resources:** Streams, rivers and reservoirs
b. **Groundwater resources:** Wells and springs
c. **Miscellaneous resources:** Rain water, run-off water and cycled water

a. Surface Water Resources

Rivers, streams, reservoirs, lakes and ponds are the important water resources for a fish farm. But, necessary permission needs to be sought for diverting the water from these resources. Rivers and streams are subjected to variations in flow and the availability in dry periods may be limited. These waters are highly contaminated and even carry high silt loads. They also contain wild or predatory fishes, parasites and disease causing organisms. Hence, these need to be filtered and disinfected before use in the fish pond. Surface waters from the areas of agriculture fields where pesticides and very harmful chemicals are used should not be diverted into the ponds. Water can be supplied by feeder canal, storage tank on pipe line by gravity or by pumping to the ponds. The most economical method is by gravity flow through earthen canals. In case of a pond constructed in a suitable soil, the minimum water supply should be 5L/sec/ha of pond throughout the year.

The canals may be of rectangular or trapezoidal. In rectangular canals, silts deposit and microorganisms harbor in the corners. These microorganisms decompose the organic matter present in the water and cause pollution. Hence, trapezoidal canals are preferred. The trapezoidal canals have the advantage of carrying more water in comparison to rectangular canals. The volume of discharge through a canal is

$$Q = A \times V$$

where,

Q = Volume of water discharge, m^3/sec

A = Cross sectional area, m^2

V = Velocity, m/sec

In case of earthen canal, velocity should be 0.6 to 0.7 m/sec. If it is less than this, the silt particles get deposited in the canal and the water velocity will be reduced. That velocity of water is called 'silting velocity'. On the other hand, if the water velocity in the earthen canal is more than 0.6 to 0.7 m/sec, then there will be formation of pits in the canal bed called 'scores'. The scores will become deeper and deeper and ultimately the canal will break. That water velocity in the earthen canal is called 'scoring velocity'. In case of closed PVC lined pipes, the water velocity can be raised to 1.0 to 1.5 m/sec.

Design of Canals

All the canals should be well designed to have the required water carrying capacity. The canals are designed using formulae that relate the carrying capacity of the canal to its shape, its effective gradient, or head loss and the roughness of the canal sides. The most commonly used formula incorporating all these factors is the Manning's Equation.

Manning's Equation

$$V = (1/n) \times R^{2/3} \times S^{1/2}$$

where,

V = Water velocity in the canal

n = Roughness coefficient of canal sides

S = Effective slope

R = Hydraulic radius of canal

R = A/P = h/2

A Wet cross sectional area (m^2)

P = Perimeter (m)

h-Depth (m)

Greater the value of R, the greater will be the flow in the canal. The wet perimeter (P) is the length of the perimeter of the cross section actually in contact with water, not including the water top width.

Thus, it is easy to calculate the carrying capacity (m^2/sec) of any open water canal with a uniform steadily flow by using the Manning's equation with formula mentioned as below,

$$Q = A \times V = A \times (1/n) \times R^{2/3} \times S^{1/2}$$

where,

A = Wet cross sectional area

V = Q/A

S = $(nV \div R^{2/3)})$

Water losses in earthen canals result from evaporation (1 to 2 percent) and seepage (5 to 40 percent), Seepage loss varies according to type of soil.

b. Groundwater Resources

When water gets evaporated from the earth surface, it is precipitated back in the form of rain, snow, *etc.*, one part of this infiltrates into the ground, forming groundwater, second major part flows as runoff in the form of rivers, and the rest is lost as evaporation and transpiration. The process whereby rain water enters the surface strata of the soil and moves to the water table, is known as 'infiltration' and that portion of water is known as 'groundwater'. Groundwater source is the preferred water source for pisciculture than surface water. The desirable groundwater table should be 2.5 to 3.0 m from the base of the pond. If groundwater table is more than 5 m, there is every chance of percolation of water into the soil whereas if groundwater table is less than 1m, water logging is a major problem.

Advantages

- ☆ More dependable and uniform
- ☆ Free from wild and predatory fish, insect *etc.*
- ☆ Less pollution
- ☆ Constant water temperature throughout the year

Disadvantages

- ☆ Low dissolved oxygen
- ☆ May contain high concentration of dissolved iron and other metals

i) Springs

A permeable stratum capable to yield groundwater under gravity is known as 'aquifer'. These vary in depth, lateral extent and thickness. Springs generally occur when an aquifer is exposed at the ground surface through a fault or crack in an upper confining layer. Springs have all the advantages of groundwater and usually do not require pumping that saves on energy costs. Reliability of a spring must be assessed before selecting it as a main source of water for an aqua farm.

ii) Wells

Well water from either deep tube well or bore well is usually the best source of water for aquaculture. The ordinary gravity wells of 2 to 5 m diameter which are constructed to tap water from the topmost water bearing strata are known as 'non-artesian wells'. The water level in these wells will be equal to the level of the groundwater table. Such wells are known as "shallow wells" of water table wells. It is necessary to record the water table at different seasons. The fluctuation of water table should not be more than 10ft.

c. Miscellaneous Resources

i) Rain Water

Ponds which rely on rainfall to fill the need for water is called 'sky pond'. Rain water is a source of free water and is of good quality. It can be collected in impoundments for later use. Rain water along with well water is most suitable for aquaculture.

ii) Surface Run-off

Rainfall run-off is also one of the major sources of water in case the ponds are required to be constructed in non-irrigated areas (plains or hills). If rainfall run-off is to be used and stored in a reservoir to affect supply to fish ponds, a ratio of 10 to 15 ha of catchment area to 1 ha of pond is required. Of course the ratio may vary slightly depending on annual rainfall of the region. Thus, three major considerations in constructing fish ponds in non-irrigated and hilly areas are:

- ☆ Availability of large catchment areas
- ☆ Availability of impermeable (non-porous) soil strata below water source and pond bed and occurrence of sufficient undergroundwater at a low depth.

Thus, the surface run-off is the water from surrounding land area which runs into them. The surface run-off depends on:

- ☆ Type of catchment
- ☆ Area of catchment

The surface run-off is maximum when type of catchment is

- ☆ Rock base or hilly terrain
- ☆ Thickly wooded contour
- ☆ Where catchment has a very steep slope

The surface run-off is minimum when catchment is

- ☆ Sandy soil
- ☆ Pasteur land
- ☆ Clayey soil with moderate percent of sand

Further, the area of catchment for a particular area can be known from the irrigation department which can give an idea of the amount of surface run-off flowing to the required area.

Thus, the volume of surface run-off can be calculated as following:

$$Q = A \times I \times R$$

where,

Q = Discharge of surface flow,

A = Area of catchment, m^2

I = Intensity of rainfall, inches/h

R = Run-off coefficient

The run-off coefficient depends on,

(a) Type of catchment (for pasteur land it is 0.35 and for rocky land it is 0.85 to 1.00)

(b) Ratio of rainfall

(c) Depending upon the discharge, number of ponds, depth of pond, *etc.*

Water Quality

The physico-chemical and biological characteristics of pond water offers the most favorable conditions for the existence of fish as well as other biota which constitute essential components of the food chain.

a. Temperature

Fishes display a great variability in their tolerance to temperature on the basis of endemicity to a region. Cold water fishes can grow well in an optimum temperature range of 10 to 12°C. Indian major carps thrive well at temperatures between 20°C to 37°C. However, high temperature ranges in tropics are responsible for higher growth rate of fishes and other biota. The temperature also plays a vital role in directly influencing various stages of fish culture *viz.* spawning, hatching, nursery management and stocking.

b. Turbidity

The water is said to be "turbid' when it contains suspended particles of silt, or clay, plankton, *etc.* It affects productivity and fish life. Decreased light penetration reduces primary production which in turn affects secondary production. Although highly turbid waters are undesirable for fish ponds. Productive ponds are generally found to have slightly turbid water. Studies on warm water fishes have shown that fishes did not show any behavioral reaction until the turbidity approached 20,000ppm (as SiO). However, the experimental fishes died at turbidity of 175,000 to 225,000 ppm due to gill choking.

The adverse effects of turbidity are:

- Decrease in light penetration
- Suspended particles adsorb considerable amounts of nutrient elements (N, P and K, *etc.*) making them unavailable for production of plankton and other vegetation
- Turbidity influences the temperature conditions of the water mass by restricting the penetration of sunlight and scattering it
- Oxygen deficiency has been reported as a result of sudden increase in turbidity
- Gill choking occurs to fish seeds

If the site selected falls in the area where water is highly turbid and the fish species to be cultured is unable to tolerate it, suitable methods of reducing turbidity have to be adopted. The use of settling tanks, different types of filters and repeated application of Gypsum (200 kg/1000m^3 initially, followed by 50 g per 1000 m^3, if necessary) have been recommended.

c. Light and Transparency

Transparency of pond water can be measured by means of "secchi disc' and the values between 25 to 70 cm are good for fish ponds.

d. pH

The most suitable pH of water for fish culture is considered to lie between 7.0 to 8.5. Water of low pH are very common in freshwater areas with soils low in calcium and rich in humic acid. Acidic water (pH 5.0 to 5.5) can be harmful to eggs, fry and adults. The most common method of correcting the low pH (acidity) is by buffering action of the lime *i.e.*'liming'. Generally it is applied in the following forms:

- Lime stone ($CaCO_3$)
- Quick lime (CaO)
- Slaked lime or caustic lime or hydrated lime [$Ca(OH)_2$].

Liming of the pond serves the following purposes in the pond:

- To correct and maintain pH of water (buffering action)
- To act as a good fertilizer as it is supplied one of the essential nutrients, the calcium
- To act as disinfectant
- To counteract the poisonous effects of excess magnesium, potassium and sodium ions
- To fix humic acid and inorganic (HSO) acid.

e. Total Alkalinity (Carbonates and bicarbonates)

Most alkalinity in natural waters is caused due to dissolution of CO_2 in water, resulting in the formation of carbonates and bicarbonates of cations of Na, Ca, Mg. K, NH, and Fe. Calcium forms the major constituent among the carbonates and bicarbonates. Carbonic acid formed after dissolution of CO_2 gets dissociated into bicarbonates (HCO_2) and carbonates (CO_2) ions in two stages. When there is an increase in pH it is indicative of increasing proportions of CO_2. At different pH values, the proportions of HCO_3^-, CO_2^- are as follows.

1. pH 0.0 to 6.35 = Carbonic acid predominates
2. pH 6.35 to 10.33 = Bicarbonate ions predominant
3. pH 10.33 to14 = Carbonate ions predominant

In highly productive waters, the alkalinity ought to be over 100 ppm. In aquaculture systems, the alkalinity should be between 80 to 200 ppm in freshwater. Alkalinity above 300 ppm adversely affects the spawning and hatching of carps.

f. Total Hardness

Hardness is the total soluble calcium and magnesium salts present in water and expressed as its $CaCO_3$, equivalent. Sulphates and chlorides of Ca and Mg are also included in total hardness. Hardness is called temporary; if it is caused by bicarbonates and carbonate salts of cations where as permanent hardness is caused by sulphates and chlorides of the metals.

Pond waters having hardness of 15 ppm or above are satisfactory for growth of fish and do not require addition of lime. Waters of 5 ppm $CaCO_3$, equivalent cause slow growth, distress and eventual death of fish.

g. Carbon Dioxide (CO_2)

When carbon dioxide content in water increases during early hours (pre-dawn), dissolved oxygen content decreases and the fishes are subjected to stress conditions. High levels of free carbon dioxide (>16 ppm) lead to poor spawning.

h. Dissolved Oxygen (O_2)

The determination of DO in the aquatic environment is among the most important factors in the water quality management in aqua farms. During daylight, DO content increases due to photosynthetic activity of plants where as at night it decreases due to its consumption by both plants and animals in water. So, during early morning hours, depletion of DO is very common in summer seasons.

Besides, extended periods of cloudy weather, destruction of vegetation or collapse of phytoplankton bloom causes depletion of DO. The water of fish ponds should possess DO, at least above 5 ppm. Fishes suffer from stress condition when DO content is around 1 to 2 ppm. At DO_2 content of 0.3 to 0.5 ppm, fishes die. Oxygen depletion can be controlled by the following methods.

- ☆ Exchange of water
- ☆ Artificial aeration
- ☆ Spraying of water on the surface of ponds or agitation of surface water
- ☆ Spraying of 5 ppm $KMnO_4$ solution over the pond surface

i. Nutrients

As a constituent of protein, nitrogen occupies the foremost rank in aquatic environment

- ☆ The dissolved nitrogen content between 0.2 to 0.5 ppm is considered favorable.
- ☆ Highly productive fish ponds should have phosphorus concentration above 0.20 ppm.
- ☆ Ratio of nitrogen to phosphorus (N:P) in water (2.6 to 1.3) though not strictly optimum, when further decreases, the productivity of pond increases
- ☆ Calcium and magnesium form the other most abundant ions in freshwater forming about 48 percent and 14 percent respectively of the total cations present in water containing 50ppm of dissolved solids

Besides, sodium, sulphur, silica and certain trace elements *viz.* waters containing 50 ppm of dissolved solids. Manganese, copper, zinc and cobalt which though occurring in extremely small amounts, are of great significance in imparting productivity to water.

B. SOIL

The second important part of site selection is the soil of the area. The two main functions of soil are:

- ☆ Ability of soil to hold water, and
- ☆ Ability of soil to provide nutrients

Based on the texture, there are six types of soil separates, *viz.*

(a) Gravel
(b) Sand
(c) Silt
(d) Clay
(e) Loam and
(f) Peat

An admixture of sand, silt and clay are necessary for the stability of soil. The soil must have enough clay to make the pond water tight. Generally impervious and quite heavy type soils are best suited for fish pond construction. Porous (sandy or gravelly) and peaty (organic) soil must be avoided as these will neither retain water nor permit compaction. From engineering point of view, soil is of two types *viz.,*

(i) Cohesive soil and

(ii) Non-cohesive soil

Cohesive soil is required for fish pond and dyke construction. This has capacity to prevent potential seepage loss from ponds and canals thereby stability and performance of dykes surrounding ponds and feeder canal can be well maintained preferably at a low cost investment. Thus, preferable soil types for pond construction are clay loam, silty loam and sandy clays. Generally soils from 2.0 m depth from the proposed pond bottom of a suitable site should have a coefficient of permeability, K = 5 x 10 m/sec. Therefore, soil investigation is to be conducted for selection of site. For this soil samples need to be collected from test pits (0.8 m x 1.5 m) of 20 m deep selected at random. In general, one pit may be done for one hectare of land. Digging a test pit permits visual examination of soil. The depth of the test pit may vary according to the elevation of water table, topography and depth of excavation. After collecting the soil samples, soil texture is to be determined by sieve method followed by bouyoucos hydrometer or international pipette method. When the relative proportion of sand, silt and clay contents of the soil are determined, the soil class/type can be known through soil textural triangle.

Besides, a general and convenient field test is to take a handful of moist soil from the test pit and to compress it into a firm ball. If the ball does not crumble after a little handling, soil contains desired clay for the purpose. To find out the water holding character *i.e.* plasticity or cohesiveness of a soil, laboratory tests may be carried out. However, mixed soils containing 20 to 30 percent clay is most desirable for fish pond construction.

Soil Fertility

Soil fertility is the measure of nutrients in the pond. The soil characteristics like texture, composition and fertility govern the water quality and pond productivity. A satisfactory pond bottom soil is that in which mineralization of organic matter takes place rapidly, nutrients are absorbed, held and released slowly over a long period. Nitrogen, phosphorus and potassium are important to the productivity of the soil and in turn of the fish pond.

a. Nitrogen

The nitrogen available to the soil is due to decomposition of organic matter by the micro-organisms through nitrogen cycle. A number of bacteria *viz. Nitrosomonas* and *Nitrobacter* found in the soil and water convert ammonium ion to nitrates by nitrification process. The nitrates are again reduced to molecular nitrogen through a process of denitrification.

In general, available nitrogen between 50 to 75mg/100 g soil is preferred for better fish production.

b. Phosphorus

Available phosphorus content in soil should be above 60 ppm for higher fish production.

c. Potassium

The optimal concentration of potassium in the soil is not well understood. However, ponds having good quality soil are usually potassium rich.

As far as organic carbon content and C/N ratio of the pond soil and its correlation with fish production is concerned, 1.5 to 2.5 percent of organic carbon and C/N ratio of 10 to 15 are adjudged to be ideal for good soil fertility. The calcium content of the soil is also considered to be important because of its relation to translocation of carbohydrate, increasing the availability of other ions, reducing toxic effects of salt solutions, forming integral component of plant tissues *etc.*

d. pH

The pH (Potentia Hydrogenii) of soil is also important, as it influence. The organic transformations of soluble phosphate and controls the absorbs and release of ions of essential nutrients at soil water interface. In pond bottom, when oxygen supply falls short, the decomposition of organic will be slow and the products of decomposition are either partially oxidized reduced to hydrogen sulphide (H_2S), methane, or short chain of fatty acids *etc.* Such compounds make the soil strongly acidic. This situation reduces the rate of bacterial action, thereby reducing the release of nutrients and, hence ultimately, reducing the productivity. Therefore, soil pH should be preferably between 6 to 9 and the ideal pH range is 7.0 to 8.5.

C. TOPOGRAPHY

Topography or the surface features of the area proposed for fish culture ponds construction is one of the most important criteria while designing the ponds. It refers to the changes in the surface elevation of natural ground whether the ground is flat, sloping, undulating or hilly. The best area for pond construction is where the ground is slightly slope. The optimum slope is between 5 to 1.0 percent. The ground slope should not be more than 2 percent. Thus, while selecting as for aquaculture venture, preference should be given to the locations where be gravitational flow may be used to fill and drain the ponds as gravity flow is economic The pond bottom should be at a higher elevation than the water table when the p is drained. Contour survey determines the elevation of the proposed site.

Sites with excessive undulating topography should be avoided as a lot of sand fill which will increase the cost of construction. Areas covered with large number of big trees and thick vegetation should be avoided as cleaning of site will be difficult and will increase cost and there is every chance of increase in seepage. Thus, from economic point of view, a pond should be located where largest storage volume can be obtained with the least amount of excavation or earth fill.

Topography can determine the kinds of pond which can be built based on whether the site is of low land or upland. Besides, cross section surveys for embankments and spill way locations are also to be undertaken carefully.

D. METEOROLOGICAL FACTORS

Among the meteorological factors which need to be considered while selecting a suitable site for fish pond construction, temperature, rainfall, wind speed and flood are important.

- ✰ Optimum temperature of 27°C is found to be most suitable for fish to grow normally. Very high temperature is not desirable. At the optimum temperature of water, dissolved oxygen content will also be maximum (9 mg/L).
- ✰ Study of distribution of rainfall pattern will give an idea of storage capacity of pond. Based on rainfall, some structures like sluice, diversion canals are to be decided to remove excess water.
- ✰ Velocity and direction of wind has direct influence while selecting a suitable site for fish pond construction. If the wind speed is in the direction of length of pond, maximum aeration in the pond can be achieved. The most desirable wind speed in the selected site should be 6.5 km/h. If the wind speed is more in a particular area already selected for fish farm, the pond should be in no case constructed along the direction of the wind.
- ✰ Generally flood inundated areas should be avoided for construction of fish farm. Based on the flood situation, height of peripheral dyke needs to be determined.

Selection of Site for Aquafarm

Stages of Site Selection

Selection of suitable site for an aqua farm can be done in two stages.They are

- ✰ Reconnaissance survey (Pre-investment Survey)
- ✰ Detailed survey

Reconnaissance Survey/Pre-investment Survey

While conducting reconnaissance or pre-investment survey, the following important points are to be considered.

- ✰ Accessibility
- ✰ Physical features of the land
- ✰ Soil characteristics
- ✰ Water source

Accessibility

The proposed site should be as far as possible near to the good, approachable road and market. During construction of farm it is easy to transport the construction materials to the site. It reduces the cost of construction and time. During culture it is easy to transport feed and fertilizer to the farm and also fish and its products to the market conveniently. It reduces the cost of production, which helps to increase the profit.

Physical Features of the Land

Physical features of the land are classified into two types.

i) Natural feature
ii) Artificial features

Natural features such as river, mountain, spring, forest, lakes, streams, rocks *etc.*,. The artificial features are ponds, buildings, electrical pole, roads, rails *etc.* It is necessary to note down all these things and prepare a rough sketch.

Detailed Survey

Main Factors – Site Condition

The proposed site for pond construction should be essentially having sufficient area to accommodate all the desired number of ponds. The site is considered to be ideal for pond construction if the land surface is flat or has uniform gentle slope in one direction. It gives facilities for quicker filling and emptying of the pond water. From the economic point of view a pond should be located where the largest storage volume can be obtained with the least amount of earth fill. This condition generally will occur at a site where the valley is narrow, side slopes are relatively steep and slope of the valley floor will permit a large deep basin.

a) Biological and Operational Factors

Before a site is selected for fish culture operation, the following factors also need to be considered.

- ✰ Species to be cultured and availability of stocking material
- ✰ System of culture (whether extensive, semi-intensive or intensive)
- ✰ Production target
- ✰ Estimated size and area required

b) Economic Factors

i) Land

- ✰ Cost
- ✰ Restrictions, if any
- ✰ Zoning regulation

ii) Water

- ☆ Availability-accessibility
- ☆ Water rights

iii) Labour

- ☆ Wages
- ☆ Availability of labour
- ☆ Liability
- ☆ Union rules
- ☆ Book keeping

iv) Transportation

- ☆ Accessibilities to facilities
- ☆ Transport to market (time required)
- ☆ Transport and handling facilities

v) Materials and Services

- ☆ Feed
- ☆ Fuel
- ☆ Equipment (pump, aerator)
- ☆ Service – maintenance
- ☆ Harvesting and packaging

vi) Laboratories

- ☆ Construction
- ☆ Earth moving (piping, valves and pumps)
- ☆ Wells
- ☆ Building
- ☆ Roads

vii) Communication

- ☆ Telephone
- ☆ Cable

viii) Power

- ☆ Availability of electricity and unit per cost
- ☆ Generator

ix) Markets

x) Taxes

c) Social Factors

i) Community Services

- ☆ Schools
- ☆ Medical care
- ☆ Housing
- ☆ Protection
- ☆ Cultural resource

ii) Environmental Impact

- ☆ Waste disposal
- ☆ Salination of adjacent land

d) Legal Factors

- ☆ Local, state and federal permit requirements
- ☆ Limitations
- ☆ Water rights
- ☆ Incentives to investors

Chapter 2

Land Survey for Aquaculture Farms

Definition

Surveying is the art of determining the relative positions of distinctive features on the surface of the earth or beneath the surface of the earth, by means of measurements of distances, directions and elevations.

Levelling is the branch of surveying which deals with the measurements of relative heights of different points on the surface of the earth. Main object of surveying is the preparation of plans and maps of the areas. This is the first step for the execution of any project.

A. Classification of Survey

1. Based on the Nature of Field

i) **Land survey**: A Land survey carried out to depict the topography of the terrain is Topographic survey. Land surveys carried out for fixing the property line, calculation of area of landed properties and preparation of revenue map of states is called Cadastral survey. Land surveys carried out for construction of roads, parks, water supply system, constructional work for any development of township *etc.* is called City survey.

ii) **Hydrographic survey**: This type of survey deals with the mapping of large water bodies for the purpose of navigation, construction of harbour works, prediction of tides and determination of mean sea level (MSL).

iii) **Astronomical survey:** It deals with determining the absolute locations of places *viz.* latitudes of different places and the direction of any line of the earth surface by making observations to heavenly bodies.

2. Based on the Purpose of the Survey

i) **Engineer's survey:** It is for getting sufficient data for designing engineering works.

ii) **Military or Defence survey:** It is for the preparation of maps of the areas of military importance.

iii) **Mine survey:** It is for the exploration of mineral wealth.

iv) **Geological survey:** It is to ascertain the composition of earth crust *i.e.*different strata of rocks of the earth crust.

v) **Archaeological survey:** It is to prepare maps of ancient culture.

3. Based on the Instruments IUsed

- ☆ Chain surveying
- ☆ Compass surveying
- ☆ Plane table surveying
- ☆ Theodolite surveying
- ☆ Tacheometric surveying
- ☆ Triangulation surveying
- ☆ Aerial surveying
- ☆ Photogrammetric surveying

i) **Plan**: A plan is the graphical representation of the features on earth surface or below the earth surface as projected on a horizontal plane. Here, generally horizontal distances and directions are shown.

ii) **Map:** The representation of the earth surface on a small scale is called a map.This must show the geographical position on the globe.

iii) **Scales:** The scale of a map may be defined as the fixed proportion which every distance between the locations of the points on the map, bears to the corresponding distances between their positions on the ground. Considering the actual surface dimensions, drawings are made to smaller scale of the area.

4. Numerical Scales

i) **Engineer's scale:** The scale on which 1 cm on the plan represents some whole numbers of meters on the ground. Ex: 1 cm = 5 m; 1 cm = 10 m *etc.*

ii) **Fraction scale:** The scale on which an unit or length on the plan represent some numbers of the same unit of length on the ground is known as fraction scale. It is also called as Representative Fraction (R.F).

Example: 1:500; To convert an engineer's scale to fraction scale, the whole numbers of meters 1: 1000 *etc.* are to be multiplied by 100. *e.g.* 1 cm = 4 m! 1: 400; 1: 500 – 1 cm = 5 m.

B. Stages of Survey Operations

The entire work of a survey operation may be divided into 3 distinct stages.

1. Field Work

It consists of 3 stages.

i) **Reconnaissance:** During reconnaissance the surveyor goes over the area to fix a number of stations along with a few permanent stations.

ii) **Observations:** Observations are obtained by the use of instruments.

iii) **Field records:** All the measurements are recorded in a field book.

2. Office Work

It consists of 3 stages.

1) **Drafting:** It refers to the preparation of plans and sections by plotting the field measurements to the desired scale.

2) **Computing:** It refers to calculating data necessary for plotting and also includes determining the areas and volumes of earth work.

3) **Designing**: It refers to selection of best alignments of roads, railways, canals *etc.* on the plotted plans.

3. Measuring the Distance

A distance can be measured by different means. Even by means of hand, a distance can be measured though not accurate. Ex. One hand: 18 inches, one step: 70 cm. However, there are two main methods of determining the distances between points on the surface of the earth. Each method has its own limitation of the degree of accuracy.

i) **Direct method:** In this method, distances are actually measured on the surface of the earth by means of chains, tapes, *etc.*

ii) **Computative method:** In this method, distances are determined by calculation as in tacheometry, triangulation, *etc.*

Instruments for Measuring Distances

The instruments used for measuring distances are as discussed below:

1. Tapes

Depending upon the material, tapes are classified as under.

i) Cloth or Linen Tape

Linen tapes are closely woven linen and varnished to resist moisture. They are generally 10 m to 30 m in length with 12 mm to 15 mm width. These are generally used to measure offset measurements and rarely used for making accurate measurements due to the following reasons.

- Easily affected by moisture and gets shrunk.
- Length gets altered by stretching.
- Not strong as chain

ii) Metallic Tape

A linen tape reinforced with brass or copper wires to prevent stretching or twisting of fibers is called a metallic tape. It is available in various lengths but tapes of 20 m and 30 m are common. It can be used for measuring accurate distances.

iii) Steel Tape

It is superior than metallic tape. Steel tapes which consist of a light strip of width 6 to 10 mm are accurately graduated. It is available in different lengths but 10 m, 20 m, 30 m and 50 m are usually used for survey measurements.

iv) Invar Tape

Those are made of an alloy of nickel (36 percent) and steel (64 percent) having very low coefficient of thermal expansion. These are 6 mm wide and are available in lengths of 30 m, 50 m and 100 m.

2. Chain

The chains used in surveying are generally of the following types.

- **Guntur's Chain**: It is 66 ft long and is divided into 100 links. Each link measures 0.66ft.
- **Engineer's Chain:** It is 100 ft long and is divided into 100 links. Each link measures 1ft.
- **Metric Chain**: It is 10 m, 20 m or 30 m long and is divided into 50, 100 or 150 links respectively. Each link measures 20 cm.
    - The length of a chain is measured from the outside of one handle to the outside of the other.
    - The length of a link is the distance between the centres of the two consecutive middle rings.

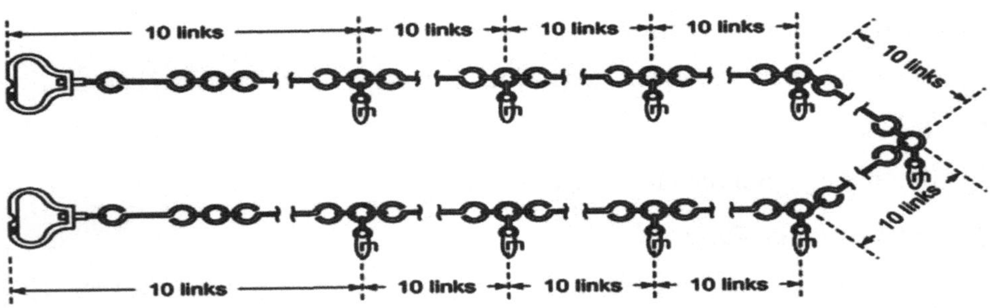

Figure 1: Metric Chain

- Metallic tags of different patterns are fixed at various important points of a chain *i.e.* 5 m, 10 m, 15 m, and 20 m for quick and easy reading of the chain.

3. **Chain Pins (Arrows)**

- An arrow is made of steel wire of 4 mm diameter.The length may vary from 25 cm to 50 cm but the length of 40 cm is of common use.
- One end of the arrow is bent into a loop of a circle of 50mm diameter and the other end is made pointed sharp. 10 arrows generally accompany a chain.

4. **Ranging Rod**

They are used for marking the positions of stations while ranging a line. It is made of well seasoned straight grained timber of teak or deodar and is generally available in 2 m or 3 m length with a 3 cm diameter. It is divided into equal parts, each part measuring 0.2 m. It is generally painted alternatively black and white throughout its length.

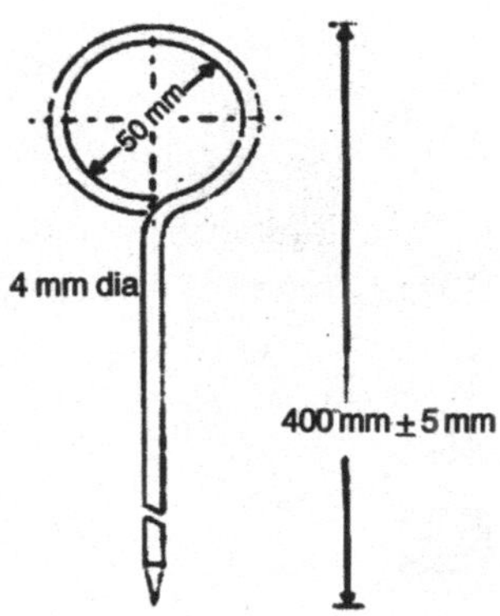

Figure 2: Chain Pin/Arrow

Figure 3: Ranging Rod

1. Chain Surveying

It is one of the methods of land surveying. It is the system of surveying in which sides of various triangles are measured directly in the field and no angular measurements are taken. It is mainly used for surveying small fish farm.

Purpose of the Chain Surveying

- To secure necessary data for exact description of the boundaries of a plot of land
- To determine the area of a plot of land.
- To prepare an accurate plan of a plot of land

Suitability of Chain Surveying

- When the ground is fairly level and open.
- When large scale plans are required.
- When the area is comparatively small in extent.

Principle of Chain Surveying

The principle of chain surveying is to divide the area into a number of triangles of suitable sides. As a triangle is the only simple plane geometrical figure which can be plotted with the lengths of its sides alone, a net work of triangles is preferred to in chain surveying.

Instruments Used for Chain Surveying

- Chain and tape
- Arrows
- Ranging rods
- Pegs
- Plumb bob
- Cross staff

i) Plumb Bob

It is used to transfer die end points of the chain onto ground while measuring distances in a hilly terrain It is also used for testing the verticality of ranging rods.

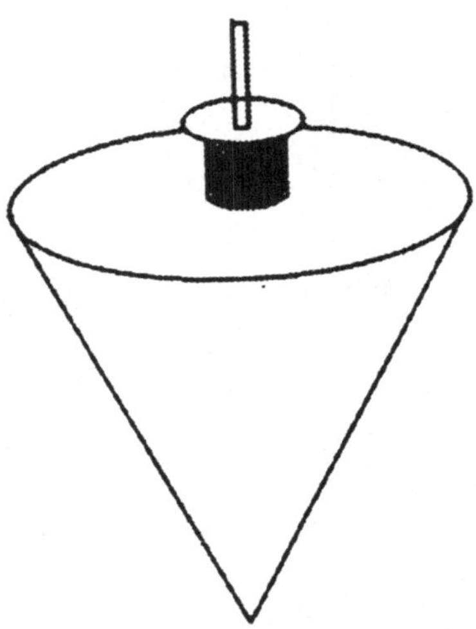

Figure 4: Plumb Bob

ii) Cross Staff

It consists of a head and a leg. The head is a wooden block octagonal or round, about 15 cm side or 20 cm diameter and 4 cm deep. The wooden block is provided with two cuts 1 cm deep at night angles to each other, establishing two lines of sight. It may be used for finding out the foot of a perpendicular offer and for setting out a right angle at a point on a chain.

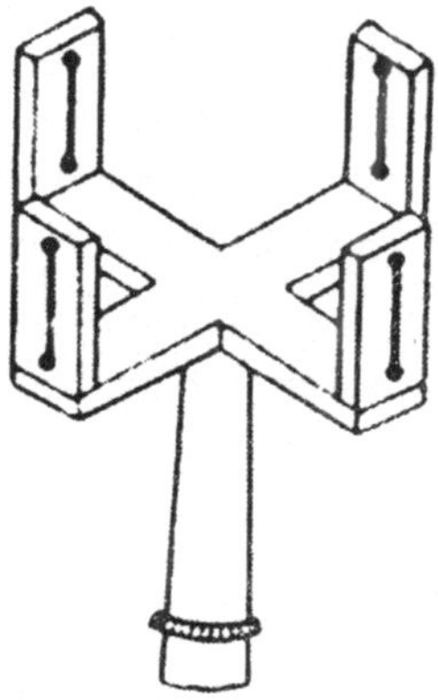

Figure 5: Cross Staff

Chain Surveying – Steps Involved

A chain survey may be done in the following steps:

i) **Reconnaissance:** The first principle of any type of surveying is to work from whole to part. Before starting the survey works, the surveyor should walk around the field and examine the ground with a view to determine how he may best arrange the work referring the buildings, roads, streams, *etc.*

ii) **Marking and fixing survey lines**: Survey stations should be marked to enable them to be easily discovered during the progress at the survey.

iii) **Running survey lines:** After completing the preliminary work, the chaining may be started from the base line. The work in running a survey line is of two fold, (i) to chain line, and (ii) to locate the adjacent details. off sets should be taken in order of their chaining. To do this, the chain is stretched along the line on the ground. Off sets are taken and then the offset distances. The point must be pulled forward. The process of chaining along with the survey line at which fences, streams, roads, *etc.* and intersected by it must be also recorded.

iv) **Plotting of a chain survey:** Generally, the scale of plotting a survey is decided before the survey is started. In general, the scale depends on the purpose of survey. The plan must be oriented on the sheet that the north side of the survey lays towards the top of the sheet and it is vertically placed by means of T-square, set square, protractor, french curves, drawing pins, divider, colour brushes. The plotting is started both sides of the line. Different Features on the ground are represented by different symbols.

2. Compass Surveying

Definition

The branch of surveying in which directions of survey lines are determined by a compass and their lengths by chaining or taping directly on the surface of the earth, is called compass surveying. If the area is comparatively large, with undulations, compass surveying is adopted. Before recommending the compass survey for any area, it must be ascertained that area is not magnetically disturbed.

A series of connected straight lines each joining two points on the ground, is called a traverse. End points are known as traverse stations and Straight lines between two consecutive stations are called traverse legs.

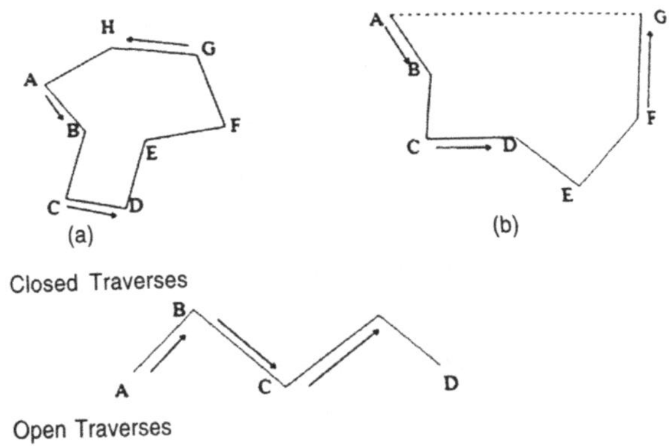

Figure 6: Closed and Open Traverses

Surveying Compass

These are of two types:

- ☆ Prismatic compass
- ☆ Surveyor's compass

Prismatic Compass

- ☆ Prismatic compass is the most convenient and portable form of magnetic compass. It can either be used as a hand instrument or can be fitted on a tripod.
- ☆ There is a magnetic needle attached to the circular ring or compass card which is made up of aluminium, a non - magnetic substance.
- ☆ When the needle is on the pivot, it will orient itself in the magnetic Meridian.
- ☆ The line of sight is defined by the object vane and the eye slit which are attached to the compass box.

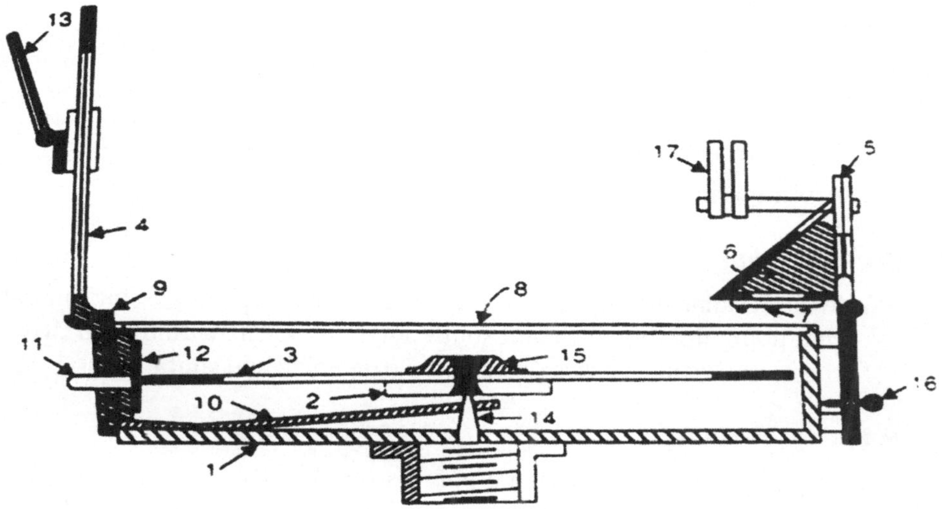

1. Box
2. Needle
3. Graduated ring
4. Object vane
5. Eye vane
6. Prism
7. Prism cap
8. Glass cover
9. Lifting pin
10. Lifting lever
11. Brake pin
12. Spring brake
13. Pivot
14. Pivot
15. Agate cap
16. Focussing stud
17. Sun glass

Figure 7: Prismatic Compass

- ☆ The object vane consist of a vertical hair attached to a suitable frame while the eye slit consist of a vertical slit cut into the upper assembly of the prism unit, both being hinged to the box.
- ☆ When an object is sighted the sight vanes will rotate with respect to the North South end of ring through an angle which the line makes with magnetic meridian.
- ☆ A triangular prism is fitted below the eyes slit.
- ☆ The prism has both horizontal and vertical faces convex, so that magnified image of the ring graduation is formed.
- ☆ The 0° or 360° reading is therefore engraved on the south end of the ring, so that bearing of magnetic meridian is read as '0' with help of prism which is vertically above south end in this particular position.
- ☆ The readings increase in clockwise direction from 0° at South end to 90° at the West end, 180° at North and 270° at East end.

Advantage of Prismatic Compass

- ☆ Both sighting the object as well as reading circle can be done simultaneously without changing the eye position.

Meridian

A meridian is any direction. There are three meridians

- ✰ True meridian
- ✰ Magnetic meridian
- ✰ Arbitrary meridian

i) *True Meridian*

- ✰ True meridian through a point is the line, in which a plane passing through that point and the north and south poles, intersects with the surface of the earth. It, thus passes through the true north and south.

ii) *Magnetic Meridian*

- ✰ Magnetic meridian through a point is the direction shown by a freely flatting and balanced magnetic needle free from all other attractive forces. The direction of magnetic meridian can be established with the help of a magnetic compass.

iii) *Arbitrary Meridian*

- ✰ Arbitrary meridian is any convenient direction towards prominent mark or signal, such as church spire or top of a chimney.

Designation of Bearings

Bearings of survey lines are designated in the following systems.

- ✰ The whole circle bearing system (W.C.B.)
- ✰ The quadrantal bearing system (Q.B.)

Whole Circle Bearing System and Reduced Bearing System

The common systems of notation of bearings are

- ✰ The Whole Circle Bearing system (W.C.B) or Azimuthal system.
- ✰ The Quadrantal bearing system or Reduced bearing system.

The Whole Circle Bearing (WCB)

- ✰ The whole circle bearing (WCB) is defined as the bearing of the line at any point which is measured with respect to the meridian is known as Whole circle bearing.
- ✰ The whole circle bearings values range from 0° to 360°.
- ✰ The whole circle bearing is generally used in the Prismatic compass.

- ☆ The whole circle bearing is used to measure the angle in the clockwise direction from the magnetic North.
- ☆ The Prismatic compass is graduated by whole circle bearing.

Quadrantal Bearing

- ☆ The **Quadrantal bearing** is also known as a **Reduced bearing**.
- ☆ Quadrantal bearings are generally measured from the North or South direction towards the East or West direction.
- ☆ The quadrantal bearing or reduced bearing can be measured either in a clockwise or anticlockwise direction.
- ☆ The **quadrantal bearing** varies from **0° to 90°**.
- ☆ In the quadrantal bearing or reduced bearing system, the bearings are taken either from the magnetic North or the magnetic south direction. It will depend on which one is nearer to that line.
- ☆ In the quadrantal bearing system, magnetic North and magnetic South lines are considered as a reference line.

Fore and Back Bearings

Every line may be defined by two bearings, one observed at either end of the line. Both the bearings expressed in W.C.B. system differ each other by 180°. The bearing of a line in the direction of the progress of survey is called Fore Bearing (FB) while the to bearing in the opposite direction of the progress of survey, is known as Back Bearing (BB) or Reverse.

Table 1: Conversion of W.B.C into Q.BL

Case	*W.C.B. between*	*Rule of Q.B*	*Quadrant*
I	0° and 90°	W.C.B	N.E
II	90° and 180°	180°-W.C.B	S.E
III	180° and 270°	W.C.B-180°	S.W
IV	270° and 360°	360°-W.C.B	N.W

Note: When a line lies exactly either along North, South, East or West, the W.C.B. of the line is converted in the quadrantal system as follows:

If W.C.B. of a line = 0° then, Q.B. of the line is N

W.C.B. of a line = 90° then, Q.B. of the line is E 90°

W.C.B. of a line = 180° then, Q.B. of the line is S

W.C.B. of a line = 270° then, Q.B. of the line is W 90°.

The rules for the conversion of Q.B. into W.C.B. are state in Table 2.

Table 2: Conversion of Q.B. into WC.B.

Case	*R.B*	*Rule of W.C.B*	*W.C.B. between*
I	Nα E	R.B	0° and 90°
II	Sβ E	180- R.B	90° and 180°
III	Sγ W	180- R.B	180° and 270°
IV	N n–W	360- R.B	270° and 360°

Theory of Magnetic Compass Surveying

- Magnetic compass gives directly the magnetic bearing of line. The bearing may either be measured in the W.C.B system or R.B system.
- The general principles of all magnetic compass depends upon the fact that, if a long, narrow strip of steel or iron is magnetized and is suitably suspended or pivoted about a point near its centre. So that it can oscillate freely about the vertical axis, it will tend to establish if self in the magnetic meridian at the place of observation.

The most essential features of a magnetic compass are

- Magnetic needle – to establish the magnetic meridian
- A line of sight – to sight the other end of the line
- A graduated circle – either attached to the box or to needle to read the direction of the lines
- A compass box – to house above parts
- A tripod – used to support the box

Theory of Prismatic Compass Surveying

- For each survey line in the traverse, surveyors take two bearings that is fore bearing and back bearing which should exactly differ by 180° if local attraction is negligible. The name prismatic compass is given to it because it essentially consists of a prism which is used for taking observations more accurately.

Chapter 3

Land Levelling for Aquaculture Farms

Levelling

Levelling is the art of determining the relative heights of points on the surface of the earth. It is a branch of surveying, the object of which are (i) to find the elevations of given points with respect to a given or assumed datums, and (ii) to establish points at given elevation or at different elevations with respect to a given or assumed datums. The first operation is required to enable the works to be designed while the second operation is required in the setting out of all kinds of engineering works. Levelling deals with measurements of vertical plane. A contour is an imaginary line on the ground joining points of equal elevations.

Methods of Levelling

There are three methods of levelling:

i. Barometric levelling
ii. Trignometric levelling
iii. Spirit levelling

i. Barometric Levelling

In this case, the difference in elevation between two points is proportional to the difference in atmospheric pressure at these points. At a given point, the atmospheric pressure does not remain constant in the course of the day, even in the course of an hour. This method is therefore relatively inaccurate and is not used.

ii. Trignometric Levelling (Indirect Levelling)

In this case, the elevation of points are computed from the vertical angles and horizontal distances measured in the field.

iii. Spirit Levelling or Direct Levelling

In this case, vertical distances with respect to a horizontal line are used to determine the relative difference in elevation between two adjacent points.

Levelling Instruments

The instruments commonly used in direct levelling:

a) A level
b) A levelling staff
c) Chain and tape
d) Arrows
e) Ranging rods
f) Pegs

A Level

A Levelling instrument used to provide a horizontal line of sight.

A level consists of the following four parts

- ✰ A telescope – To provide line of sight
- ✰ A level tube - To make the line of sight horizontal
- ✰ A levelling head - To bring the bubble in its centre of run
- ✰ A tripod – To support the instrument

There are four types of Level

- ✰ Dumpy level
- ✰ Wye (Y) level
- ✰ Reversible level
- ✰ Tilting level

Dumpy Level

Dumpy level is mostly used for levelling survey from the engineering point view. It is a telescope rigidly fixed to its support. It can either be rotated about its longitudinal axis nor can it be removed from its support. The instrument is more stable when compared to others. It consists essentially of the following parts.

a. **Telescope**: It is an optical instrument used for magnifying and viewing the images of distant objects. In survey, it is used to provide a line of sight.

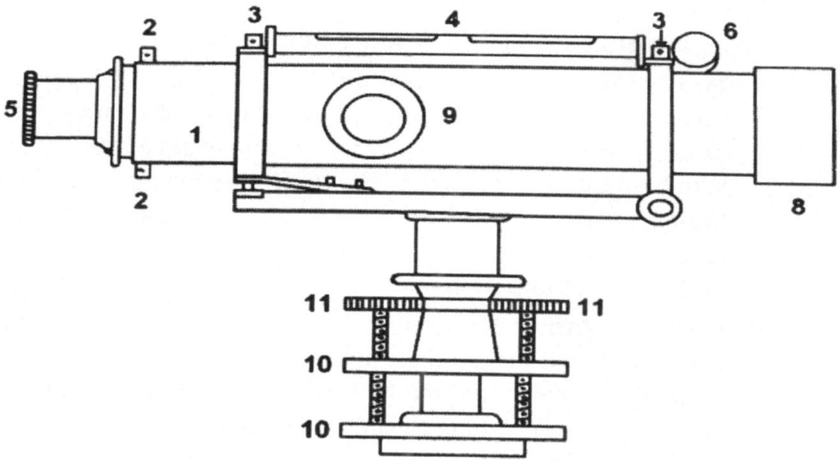

1. Telescope
2. Diaphragm adjusting screws
3. Bubble tube adjusting screws
4. Longitudinal bubble
5. Eye piece
6. Transverse bubble tube
7. Objective end
8. Ray shade
9. Focussing screws
10. Levelling Head
11. Foot screws

Figure 1: A Dumpy Level

b. **Level tube:** It is to make the line of sight horizontal. It is also known as "bubble tube", consists of a glass tube placed in a brass tube which is sealed with plaster of paris. It is nearly filled with either ether or alcohol or a mixture of both Remaining space is occupied by an air bubble. The centre of the bubble always rests at the highest point of the tube.

c. **Levelling head:** It is to bring the bubble of the tube level at centre of its run. It consists of two parallel plates with three foot screws, upper plates at known as "Tribarch" and lower as "Trivet" which can be screwed on to the tripod.

   It has mainly three functions.

   i. To support the telescope
   ii. To attach the level to the tripod
   iii. To provide a means for level

d. **Tripod:** It is to support the above three parts of the level. When in use, level is supported on a tripod which consists of three solid or framed legs provided with pointed iron shoes.

Adjustment of a Dumpy Level

a. **Setting up the level:** The instrument is fixed on the tripod.

b. **Levelling the instrument:** The instrument is levelled approximately by leg adjustment and the telescope must be placed at a convenient height.

   Two foot screws are to be held with the thumb and 1st finger of each hand, turn them uniformly so that the thumbs move either towards each other or away from each other until the bubble is at centre. The upper plate is to be rotated through 90°. The third screw needs to be held with the thumb and first finger of the right hand and turn it until the plate bubble is central. The process is to be repeated till the bubble remains central in both the positions.

c. **Levelling staff:** It is a straight, rectangular wooden rod graduated into metres or ft and smaller divisions. Bottom of the staff represents the zero reading. The reading given by the line of sight on a levelling staff held vertically is the height of the line of collimation from the point on which staff is held. This may be divided into two classes.

   i) **Self reading staff:** A staff on which readings are directly read by the observer through the telescope.

      It is of three types

      ☆ solid staff,

      ☆ Folding staff, and

      ☆ Telescopic staff

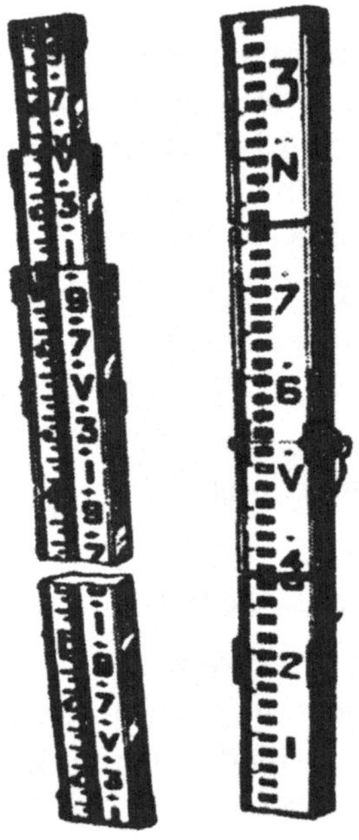

Figure 2: Levelling Staff

The telescopic staff is 4m divided into 3 portions. Bottom one is 1.5 m, middle one is 1.25 m and upper part is 1.251 m. The middle and upper parts are held by brass spring latches.

ii) Target staff: A target which can be moved up and down is attached to the staff. For taking readings, the level man directs teh staff man to raise or lower the target till it is bisected by the line of sight.

Taking a Staff Reading

a. The instrument is set up and carefully levelled by temporary adjustment.
b. The staff is held vertically over the staff station. To hold the staff vertically the staff man stands behind the levelling staff with his heels together and the bottom of the staff in between his toes and holds the staff in between the palms at the height of his face.
c. The telescope is directed towards the staff by the instrument man and it is focussed by using the focussing screw. The telescope is moved till the staff is exactly between the vertical hairs of diaphragm.
d. If the bubble on the telescope is not at the centre, the same is centred by using the foot screws.

Technical Terms used in Levelling

1. **Datum surface:** This is an arbitrary surface with respect to which the elevations of the points are measured and compared. The mean sea level (MSL) affords a convenient datum world over and elevations are commonly as so much above or below sea level.
2. **Elevation:** The elevation of a point on or near the surface of the earth is its vertical distance above or below an arbitrary assumed level surface or level. The difference in elevation between two points is the vertical distance between two level surfaces in which the two points lie.
3. **Mean sea level (MSL)**: Mean sea level is the average height of the sea for all stages of the tides. At any particular, it is derived by averaging, the hourly tide height over long period of 19 years.
4. **Line of sight:** Line passing through the optical centre of the objective, eye piece and entering the eye.
5. **Line of collimation:** Line joining the intersection of cross hairs to the optical centre of objective and its continuation.
6. **Bench mark:** Bench mark is a relatively permanent point of reference whose elevation with respect to some assumed datums is known. It is used as a starting point for levelling or as a point upon which to choose as check.
7. **Instrument station:** Point where instrument is set up for observation
8. **Station:** Point where levelling staff is held.

9. **Height of instrument:** The elevations of the line of sight with respect to the assumed datum.
10. **Back sight (B.S.):** The first staff reading taken after setting up of the instrument in any position. This will always be a reading on a point of known elevation.
11. **Fore sight (F.S.):** Last staff reading taken before moving the instrument. This will always be a reading on a point whose elevation is to be determined. This reading indicates the shifting of the instrument.
12. **Intermediate sight (L.S.)**: Any staff reading taken on a point of unknown elevation after the B.S. and before F.S. This is necessary when more than two staff readings are taken from the same position of the instrument.
13. **Change point (C.P.):** It indicates the shifting of the instrument. Both B.S.and F.S. are taken on a change point.
14. **Booking and reducing level:** This is the vertical distance above or below. the datum. It is either +ve or -ve. There are two methods of booking and reducing the elevations of points from the observed staff readings: (i) Collimation or height of instrument method, and (ii) Rise and fall method.
    a. **Collimation or height of instrument method:** In this method the height of instrument (H.I.) is calculated for each setting of the instrument by adding back sight (B.S.) to the elevation of B.M. (First point). Their elevation of reduced level of the turning point is then calculated by subtracting its foresight (F.S.) from H.I. For the next setting of the instrument, the H.I. is obtained by adding the B.S. on previous result to its R.L. The process continues till the R.L. of the last point (F.S.) is obtained by subtracting the staff reading from the height of the last setting of the instrument.

       Arithmetic check: The difference between the sum of B.S. and the F.S. should be equal to the difference between the last and the first R.L.

       Thus,

       B.S. – F.S. = Last R.L. – First R.L.

       The method affords a check for the H.I. and R.L. of turning points but not for the intermediate points.
    b. **Rise and fall method:** In rise and fall method,the height of instrument not at all calculated but the difference of level between consecutive points is found by comparing the staff readings of two points for the same setting of the instrument. The difference between their staff readings indicates a rise or fall according as the staff reading at the point is smaller or greater than that at the preceding point. The figures

for 'rise and fall' worked out thus for all the points give the vertical distance of each point in above or below the preceding one, and if the level of any one point is known, the level of the next will be obtained by adding its rise or subtracting its fall as the case may be.

Arithmetic check: The difference between the sum of B.S. and the F.S. should be equal to the difference between the sum of rise and sum of fall and should also be equal to the difference between the last and the first R.L.

Thus,

B.S.-F.S. Rise – Fall= Last R.L. – First R.L.

Theodolite Survey

The Theodolite is the most precise instrument designed for the measurement of both horizontal and vertical angles and has wide applicability in surveying such as laying of horizontal angles, determining difference in elevation, setting out curves *etc.* A transit theodolite and a levelling staff and the instruments used for chains survey are required for theodolite survey.

Plane Table Survey

Plane table surveying is the graphical construction of straight lines, angles and triangles for plotting the ground detail points. This is done for large fish farms. Here, field observations and plotting proceed simultaneously.

Principle

The principle of plane tabling is based on the fact that the lines joining the points on the plane table, are made to lie parallel to corresponding lines joining the ground points while working at each station. The principle is mainly by considering the graphical reduction of a triangle to given dimensions. The base of the triangle is plotted on the desired scale and the base angles are plotted directly by turning the alidade at each end. The intersection of the rays gives the desired location of the triangle vertex.

Instruments Used

The instruments required for plane tabling are:

- ✰ Plane table with stand
- ✰ A plain Alidade
- ✰ Spirit level
- ✰ A magnetic compass
- ✰ Plumbing fork
- ✰ Drawing paper

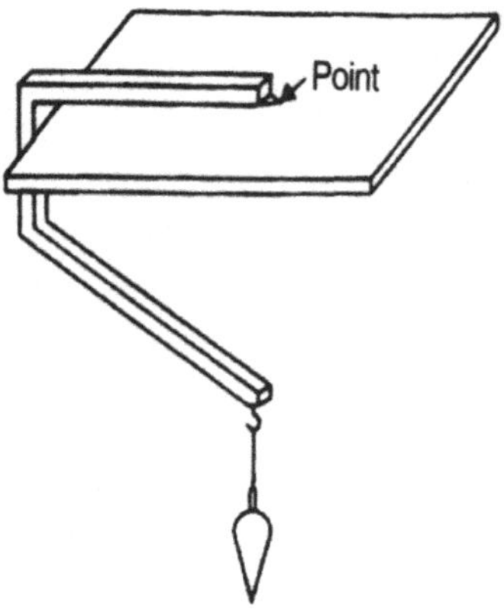

Figure 3: A Plumbing Fork

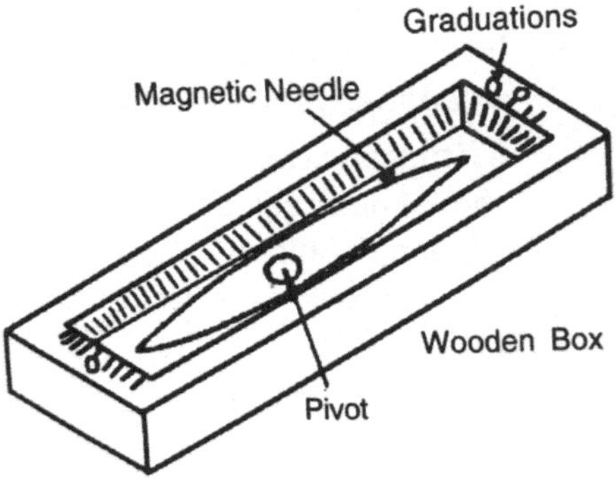

Figure 4: A Compass

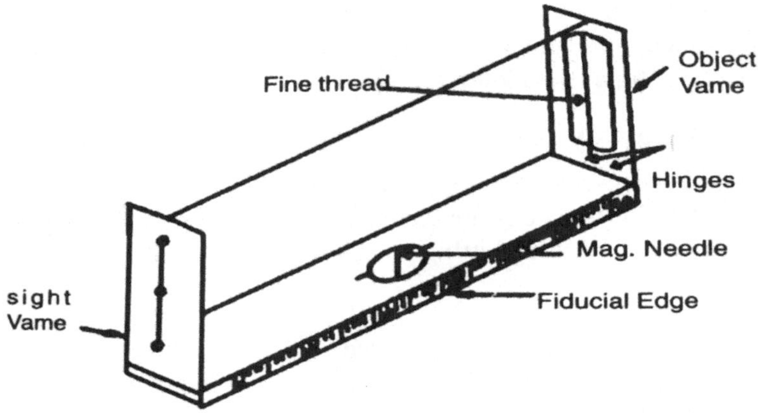

Figure 5: A Plain Alidade

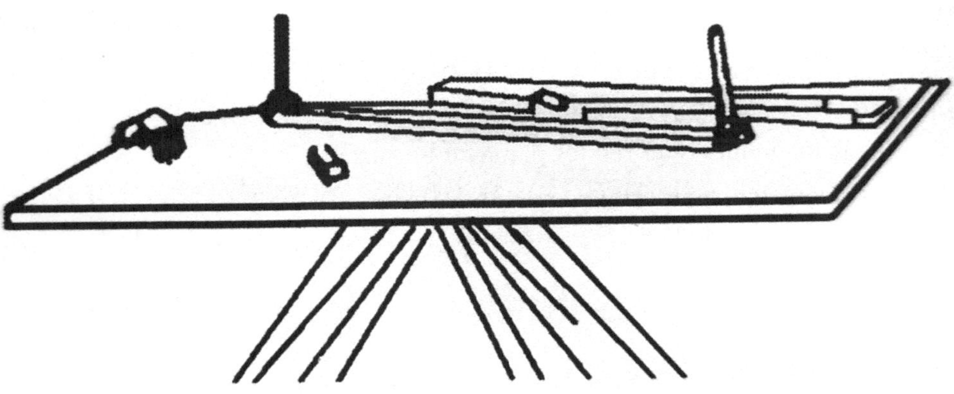

Figure 6: A Plain Table with an Alidade

Working Operations of Plane Table Surveying

The setting up of the plane table includes three operations:

- ☆ Levelling
- ☆ Orientation
- ☆ Centering.

i) Levelling

Place the table over the station-point and spread out legs to make it level and at a convenient height, preferably not above the elbow. Then oriented and centre the table approximately as explained in the following two operations and complete the final levelling. The levelling is done by means of levelling-screws or the ball and socket arrangement or simply by adjusting the legs if there is no levelling head, with reference to a circular level or a level tube placed in two positions at right angles to each other.

ii) Orientation

The table is said to be oriented when it is so placed that all lines on the paper are parallel to the corresponding lines on the ground. This is necessary when more than one stations have to be occupied by the table as otherwise the board would not be kept parallel to itself at the various stations, and the effect would be similar to using a different meridian at each station.

There are two methods of orienting the table *viz.*:

(a) By the use of a magnetic needle and

(b) By back-sighting.

iii) Centering

The point on the paper representing the station being occupied should be vertically over the point on the ground. This operation is known as centering of the table. This may be done by means of plumbing fork or U-frame. To do this, place the pointed end of the fork touching with the point on the paper and suspend a plum-bob from the other end. Shift the table bodily until the plumb bob hangs vertically over the centre of the station peg. If plumb-bob is not available, centering may be done by dropping a piece of stone from a point on the underside of the board which is directly under the point on the paper. Accurate centering is necessary for large scale maps and when the scale is small, accurate centering becomes a needless task.

Orientation and centering are inter-related because when one is done, the other gets disturbed, therefore, both of them have to be done simultaneously.

Methods of Plane Table Surveying

- ☆ There are mainly four methods of plane table surveying, radiation, intersection or triangulation, traversing, and resection.

☆ The first two methods are generally employed for locating the details, while the other two methods are used for locating the plane table station.

Contouring

Contours: An imaginary line, on the ground, joining the points of equal elevation above the assumed datum, is called a contour. It is a plan projection of the plane passing through the points of equal height on the surface of the earth. The least horizontal distance between two consecutive contours, is called horizontal equivalent and the vertical distance between any two consecutive contours is called contour interval.

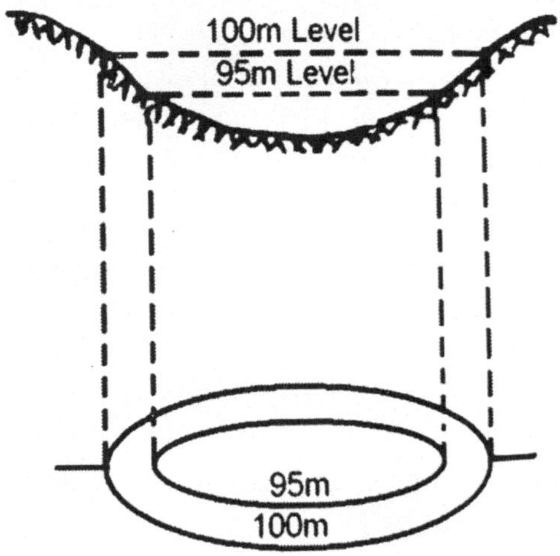

Figure 7: A Contour

Characteristics of Contour

Two contour lines of different elevation cannot cross each other. If they did, the point of intersection would have two different elevations which is absurd. Howeverm contour lines of different elevations can intersect only in the case of overhanging cliff or a cave.

Contour lines of different elevations can unite to form one line only in the case of a vertical cliff.

Contour lines close together in indicate steep slope.

(a) They indicate a gentle slope if they are far a part.

(b) If they are equally spaced uniform slope is indicated.

(c) A series of straight, parallel and equally spaced contours represent a plane surface.

☆ A contour passing through any point is perpendicular to the line of steepest slope at that point.

- ☆ The perpendicular distance between contour lines is the shortest distance.
- ☆ A closed contour line with one or more higher inside it represent a hill. Similarly a closed contour line with one or more lower ones inside it indicates depression without an outlet.
- ☆ Two contour lines having the same elevation cannot unite and continue as one line. Similarly a single contour cannot split into two lines.
- ☆ A contour line must close upon itself, though not necessarily within the limits of the map.
- ☆ Contour lines cross a water shed or ridge line at right angles.
- ☆ They form curves of U-shape round it with the concave side of the curve towards the higher ground.
- ☆ Contour lines cross a valley line at right angles. They form sharp curves of v-shape across it with convex side of the curve to words the higher ground.
- ☆ If there is a stream the contour on either side, turning upstream, may disappear in coincidence with the edge of the stream and cross underneath the water surface.

Methods of Contouring

Field work for locating contours may be executed by two methods:

(a) Direct method and

(b) Indirect method

- ☆ In the direct method, contours to be plotted are actually traced in field by locating and marking a number of points on each.
- ☆ These points are surveyed on a plane table section and the appropriate contours are drawn through them.
- ☆ This method is comparative slow and is generally not adopted on large surveys unless a superior accuracy is demanded.
- ☆ It is suitable for contouring of small areas where better accuracy is desired.

In the indirect method, sufficient numbers of points are given spot levels. The locations of such points can be conveniently plotted on a plane table section as these generally form the corners of well shaped geometrical figures i.e squares, rectangles, triangles *etc.* This method of contouring is sometime is known as 'contouring by spot levels. Indirect method of contouring can be employed in small scale surveys for extensive areas. This method is cheap quicker and less tedious as compared to direct method of contouring method of contouring can be employed in 3 different ways, *viz.* (i) By square method, (ii) By cross section method, and (iii) By tacheometric method and square method is most common.

By Square Method

In this method, the entire area is divided into a number of squares, the sides of which may vary from 5 m to 25 m, depending up the nature of the ground, the contour interval, and the scale of the plan. The squares are plotted on the desired scale of the plan and reduced levels of the stations as well as that of the salient features are entered. The contours of the desired values are then interpolated.

Suitability

This method is suitable in low undulations without any vegetative covers.

Uses of Contours

Contour maps are very useful since they provide valuable information about the terrain. Some of the uses of contours are as follows:

- ✰ The nature of the ground and its slope can be estimated
- ✰ Earth work can be estimated for construction of ponds
- ✰ Sites for road works, canals, channels, dams *etc.* Can be decided
- ✰ It is possible to identify suitable site for any project from the contour map the region
- ✰ Inter-visibility of points can be ascertained using contour maps

Computation of Areas

Various formulae such as Trapezoidal rule, Simpson's rule, Mid ordinate rule or the Average ordinate rule may be applied for computation of the area.

i) Trapezoidal Rule

The area is divided into numbers of trapezoids and the following rule is applied: The sum of the first and the last plus twice the sum of the intermediate ordinates multiplied by half the common distance between the ordinates gives the required area.

Area A = d/2 [(Oo + On) + 2(01 + 02+.............+ On-1] (1)

Here Oo, 01, 02, On are the ordinates and d is the common distance between the ordinates. If the base line cuts the boundary at one end or both ends of the figure, 00 or On or both are zero.

ii) Simpson's Rule

The rule assumes that the short lengths of the boundaries between the ordinates are parabolic areas; and the rule is, therefore, called parabolic rule. To the sum of the first and last ordinates is added twice the sum of the remaining odd ordinates and four times the sum of all the even ordinates. This total sum multiplied by one-third the common distance between the ordinates gives the required area.

Area A d/3 [Oo+On+4(01+03+05+.+ On-1) + 2(02+04+06+........+ On-2)].....(2)

The rule requires an even number of division of the area (*i.e.* and odd number of ordinates). Here, Oo, O3, O5 *etc.* are even ordinates and O2, O4, O6 *etc.* are odd.

Earth Work Calculation

Before constructing a fish farm, it is required to make plan, design and layout of the fish farm. For this, it is essential to know the estimated cost. Looking to the financial budget, the design of the farm may be changed. While preparing an estimate, the quantities of different items of works are calculated by simple mensuration method and from which estimated cost is being calculated. The calculations mainly consist of length x breadth x height. Accuracy in estimate is very important. In preparing an estimate, one has to go into details of each item. The rates in the estimate may vary from time to time and place to place. For the estimating purpose, the metric system is mainly adopted as it is very simple. For fish farm engineering works, the units which are commonly used for estimating are meter for length, square meter (m^2) for area, cubic meter (m^3) for volume, kilogram (kg) for mass and litre (L) for capacity.

Earthwork

Earthwork in pond excavation and earthwork filling in dyke construction are usually the main items of work in case of pond construction. Pond is dug to the exact length, breadth and depth of different levels. Considering the pond side slope of 2:1, the earth cut area at three different levels.Depth or, average depth of earth work in cutting, earth available from pond excavation is required to be utilized in pond dyke construction. Therefore, for economy of construction, the earthwork in excavation should be as far as practicable balance with the earth work in filling for raising dykes. Even after balancing, if there is surplus earth, the excess earth may be utilized in levelling some other work. Earth work in any type of soil in dyke filling is in layers not exceeding 30cm in depth including rough dressing breaking clods to maximum 5 to 7 cm with proper compaction by watering and ramming.

Earthworks are engineering works created through the moving and/or processing of massive quantities of soil or unformed rock. Earthwork is done to reconfigure the topography of a site to achieve the design levels.

Earthwork involves cutting and filling to achieve the required topography. Cutting is the process of excavating earth material from a work location to achieve the desired topography. Filling is the process of moving the excavated material or additional earth material to a work location to achieve the desired topography.

Longitudinal Slope and Cross Slope

The slope of the ground along a given line is known as the longitudinal slope. On the other hand slope of the ground perpendicular to the given line is known as the cross slope or transverse slope.

A ground having a cross slope is also known as side long ground. In flat ground the cross slope is usually negligible. In a hilly terrain the cross slope is usually significant. Unless otherwise specified we should always assume that cross slope is zero.

Side Slope (H : V)

For stability of a dike, side slope is required. The side slope mainly depends upon soil texture and prevailing site conditions. The flatter the slope, more stable is the dike. Ideal side slope is 1.5:1 to 2:1 (H:V).

Table 1: The Minimum Side Slope Required for different Soil Textures

Sl.No.	*Type of Soil*	*Side Slope (H:V)*
1	Clayey soil	1;1
2	Loamy soil	1.5:1
3	Sandy soil	2:1
4	Homogeneous with graded soil	2:1
5	Homogeneous coarse silts	2.5:1
6	Homogeneous clay	2:1
7	Loamy soil	2.5:1

Cross Sectional Area

When both outer and inner slopes of dikes are same

$$A=Bd + Sd^2$$

where,

A = Cross sectional area in Sq.meters

B = Top Width (Crest)

D = Depth

Slope = (H:V)

When inner and outer slopes are not same, the cross sectional area should be calculated separately as two triangles and one rectangle, and finally, areas of all three sections are added.

The quantity of earthwork required in the construction of an embankment is calculated by using following equation.

where,

$$Q = L x A$$

L = Length of an embankment.

A = Cross sectional area

Cost of Construction

Cost of construction = Rate of earthwork x Quantity of earthwork

Chapter 4

Soil

Soil is complex mixture of living organisms, organic matter, minerals, water, sand air. Considering the volumetric composition of inorganic soil mineral matter is 45 percent, organic matter 5 percent, soil water 25 percent and soil air 25 percent. Among the living organisms, macro-organisms like rodents, insects and worms and micro-organisms like bacteria, fungi and algae live in soil in large numbers. Soils (sediments) have an important effect on water quality. The pH and mineral concentration of soil have a direct effect on the acidity and presence of toxic metal ions in water supplies. Depending on their texture, structure and consistency, various kinds of soils hold more or less water and air. The layer down below the soil is called 'sub soil' which is more or less compact and less fertile.

- ☆ The term soil has various meanings depending upon the general professional field in which it is being considered.
- ☆ To an agriculturist soil is the substance existing on the earth surface, which grows and develops plant life.
- ☆ To an engineer soil are the un-aggregated or un-cemented deposits of mineral covering large portion of earth's crust.

Types of Soil

Soil Classification

The type of soil and its composition at a site has direct bearing on the productivity of a pond. The site should contain soft bottom soil or mixed soil comprising of sand, silt and clay to ensure good water holding capacity as well as production of natural food organisms on which fish feed and grow. Soil classification is normally based on particle size or some additional soil properties such as plasticity and compressibility. Based on the relative proportion of each of the three soil separates *viz.*, sand, silt and clay, *viz.*, 12 textural classes have been identified. Textural classes along with their percentage of sand, silt and clay have been presented in the Table 1.

Table 1: Twelve Textural Classes and their Percentage of Sand, Silt and Clay

Group	*Textural Class*	*Range in Percentage*		
		Sand	*Silt*	*Clay*
Coarse Textured	1. Sand	85-100	0-15	0-10
	2. Loamy sand	70-90	0-30	0-15
	3. Sandy loam	43-80	0-50	0-20
Moderately Textured	1. Loam	23-52	20-50	7-27
	2. Silty loam	0-50	50-88	0-27
	3. Silt	0-20	80-100	0-12
	4. Sandy clay loam	45-85	0-28	20-35
Fine Textured	1. Clay loam	20-45	15-53	27-40
	2. Silty clay loam	0-20	40-73	27-40
	3. Sandy Clay	45-65	0-20	35-45
	4. Silty Clay	0-20	40-60	40-60
	5. Clay	0-45	0-40	40-100

Soil Texture

It refers to the relative proportion of various size groups of particles in a given soil. Most soils are mixtures of sand, silt and clay.

a. **Sand:** These are very large sized particles and have the large size pore spaces. Their water holding capacity is low. They do not possess plasticity.
b. **Clay:** These particles are smallest in size and possess fine pore spaces. With decreasing particle size, there is decrease in rate of aeration. The water holding capacity (WHC) is very high. Property like plasticity is also very high when moist, clayey soils are exceedingly sticky and when dry, become very hard and difficult to break.
c. **Silt:** These particles are intermediate in size. These also show properties some what between sand and clay. Good loamy soils contain about 30 to 50 percent silt. In drainage,absorption of water and WHC,they are intermediate between sandy soils and clayey soils.

Quick Field Tests to Determine Soil Texture

For fish pond construction, it is better to have a soil with a high proportion of salt and/or clay which will hold water well. To check quickly on the texture of the soil at different depths, the following quick test can be done.

i) Through-the-Ball Test

- ☆ Take a handful of moist soil and squeeze it into a ball
- ☆ Throw the ball into the air about 50 cm and then catch it
- ☆ If the ball falls apart, it is poor soil with too much sand
- ☆ If the ball sticks together, it is probably good soil with enough clay in it

ii) Soil Texture by Feel Method

In the field, texture is commonly determined by 'feel' method. The soil is rubbed between thumb and fingers, preferably in the wet condition. Sand feels gritty and its particles can be easily seen with the naked eye. The silt when dry feels like flour or talcum powder and is slightly plastic when wet. Clayey material feels very plastic and exhibit stickiness when wet and are hard under dry condition. A more accurate and fundamental method is 'laboratory method' which is based on mechanical analysis.

Soil Sampling

Types of Soil Sampling

The soil samples are classified into two groups

- ☆ Disturbed sample
- ☆ Undisturbed soil samples

Types

i) Disturbed Sample

A disturbed soil sample is that in which the natural structure of soils get partly or fully modified and destroyed although with suitable precautions the natural water content may be preserved which do not represent exactly how the soil was in its natural state before sampling. Disturbed samples are used for the more simple tests that will be performed and particularly on the field.

ii) Undisturbed Soil Samples

An undisturbed sample is that in which the natural structures and properties remain preserved. This represents exactly how the soil was in its natural slate before sampling. Undisturbed samples are necessary for the more sophisticated tests which must be performed in the laboratory for more detailed, physical and chemical analysis. Undisturbed samples must be collected with the greater care for they should represent exactly the nature of the soil. For complete study of the soils in the site it is necessary to conduct both disturbed and undisturbed samples test. Soil samples for aquaculture are normally taken to a depth of 2m. So, each soil horizon up to that depth can be examined. If the water table is found at a depth of less than 2m soil samples should always be taken as deep as possible.

Soil Sampling Methods

Soil samples may be taken by three methods which involve either digging, drilling or boring holes at the place on your site where you have chosen to collect samples from different depths.

i) Open Pit Method

Open test pits are the only means available actually to see and be able to examine a soil profile in its natural state. They may be dug by hand or by special excavating equipment such as a trench digger. If necessary, you may obtain undisturbed samples from selected horizons of an open pit.

ii) Auger Boring Method

The auger boring method is a way to obtain soil samples from different depths by drilling, without having to dig a pit. This way, a continuous series of soil samples are taken which make it possible to assemble a core showing the soil horizons. The auger boring method is cheap and fast, one can quickly check the soil at several places on the site, but it provides only disturbed samples. An auger may be used in most soils above the water table and in cohesive soils below the water table.

iii) Thin-Walled Tube Method

With the thin-walled tube method, one uses a light weight tube or pipe which is open at both ends. It is pushed into the soil to obtain a sample and is then removed from the hole. If it is well done, this method will provide undisturbed samples.

Standard thin-walled tubes are manufactured about 30-60 cm long and 4-7 cm in diameter. You can also make your own tubes from lengths of steel pipes made of No. 16 (1.6 mm) or No. 18 (1.25 mm) gauge steel, 15-30 cm long and 3.5-5 cm in diameter.

Three Phase System of Soil

In general, the soil sample will be consisting of soil solids, water and air, which is designated as three phase system of soil. The sum of the air content and water content of a sample is called as voids. If the voids are completely filled with water such a sample is called as a purely saturated sample or 100 percent saturated sample. If the voids of soil is completely filled with air such a sample is called as purely dried sample, on the other hand if the part of the voids are filled with water and remaining with air it is called as a moist sample. The figure shows the diagrammatic representation of the three phase system of soil element in terms of volumes and weights respectively.

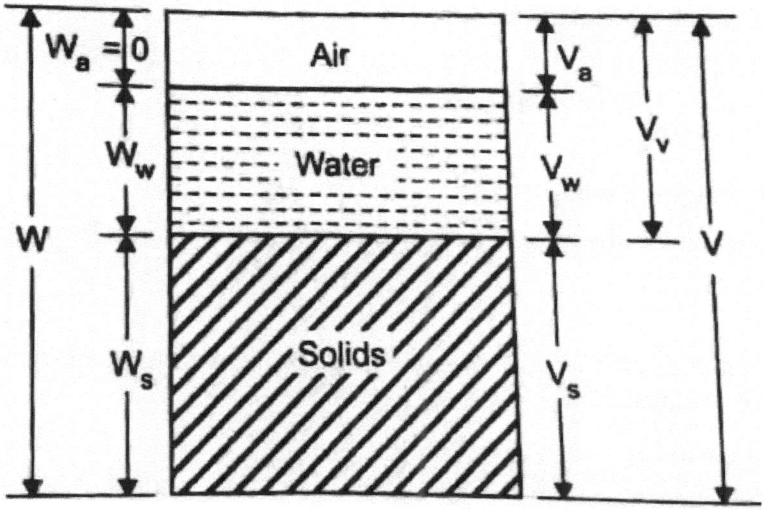

V_a: Volume of air in the sample; V_w: Volume of water in the sample; V_y: Volume of voids in the sample; V: Total volume of the soil sample; V_s: Volume of soil solids in the sample; W_a: Weight of air in the soil sample can be considered as zero; W_w: Weight of water in the soil sample; W_s: Weight of soil solids in the sample; W: Total weight of soil sample.

Further $V_v = V_a + V_w$

☆ Total volume $V = V_s + V_v$; ☆ Total weight $W = W_w + W_s$

Figure 1: Diagramatic Representation of 3 Phase System of Soil

Definition of Soil Properties

1. Water Content or Moisture Content (W)

Water content can be defined as the ratio of weight of water to the weight of solids in a given sample.

$$W = W_w/W_s$$

It is always be expressed as percentage.

2. Bulk Density or, Unit Weight (p)

It is defined as the ratio of the weight of the soil sample to its volume.

$$p = W/V$$

The unit of bulk density is g/cc, kg/m^3 or, t/m^3

3. Dry Density (pd)

Dry density can be defined as the ratio of the weight of the soil solids to the total volume of the soil sample.

$$pd = W_s/V$$

4. **Unit Weight of Soil Solids (ps)**

It can be defined as the ratio of weight of soil solids to the volume of the soil solids in a given soil sample.

$$ps = W_s/V_s$$

5. **Unit Weight of Water (pw)**

It is defined as the ratio of weight of water to the volume of water in a given soil sample.

$$pw = W_w/V_w$$

the value of unit weight of water can be taken as I g/cc or, t/m^3, unless otherwise specified.

6. **Void Ratio (e)**

It is defined as the ratio of volume of voids to the volume of soil solids in a given soil sample.

$$e = V_v/V_s$$

7. **Porosity (n) or Percentage of Voids**

Porosity is defined as the ratio of volume of voids to the total volume of the soil sample. This is expressed as percentage.

$$n = V_v/V$$

8. **Specific Gravity (G)**

Specific gravity is defined as the ratio of unit weight of soil solids to the unit weight of water.

$$G = ps/pw$$

9. **Degree of Saturation (S)**

It is defined as the ratio of volume of water to the volume of voids in the given soil sample.

$$S = V_w/V_y$$

It is expressed as percentage. If the soil sample is completely saturated then. $V_w = V_v$ or, S=1 *i.e.* V_w/V_v =1 is 100 percent. Similarly, if the soil is purely dry, V_w = 0, Then, S = 0.

Soil Permeability

It is the property of the soil to transmit water and air and is one of the most important qualities to consider for fish culture. A pond built in impermeable soil will lose little water through seepage. The more permeable the soil, the greater the seepage. Soils are generally made up of layers and soil quality often varies greatly from one layer to another. Before pond construction, it is important to determine the relative position of the permeable and impermeable layers. The design of a pond should be planned to avoid having a permeable layer at the bottom to prevent excessive water loss into the subsoil by seepage. A soil permeability of K (Coefficient of permeability) less than 5×10^{-6} m/s is desirable.

Permeability is defined as the property of a porous material which permits the passage of water through its inter-connecting voids. A material having continuous voids is called permeable. Gravels are highly permeable while stiff clay is the least permeable. Therefore, clay is used in construction of dikes of fish ponds. Soil structure is an important factor that influences the permeability. Soil structure is defined as the state of arrangement and aggregation of soil structural particles in a soil mass.

Darcy's Law

In 1856, Darcy confirmed the applicability of principles of fluid flow in capillary tubes. It states that the rate of flow or discharge per unit time is proportional to hydraulic gradient.

$$V = k \times i$$

where,

V = Velocity of flow (m/s)

K = Coefficient of permeability

I = Hydraulic gradient

Coefficient of permeability is defined as the average velocity of flow that will occur through the total cross-sectional area of soil under unit hydraulic gradient. It is expressed as m/day or cm/sec.

Factors Affecting Permeability

- Grains size
- Properties of pore fluid
- Voids ratio of the soil
- Structural arrangement of soil particles
- Entrapped air and foreign matter
- Adsorbed water

Determination of Coefficient of Permeability

It can be determined in laboratory by direct measurement with help of parameters by allowing the water to flow through soil sample either under constant head or falling head.

Soil Texture Triangle

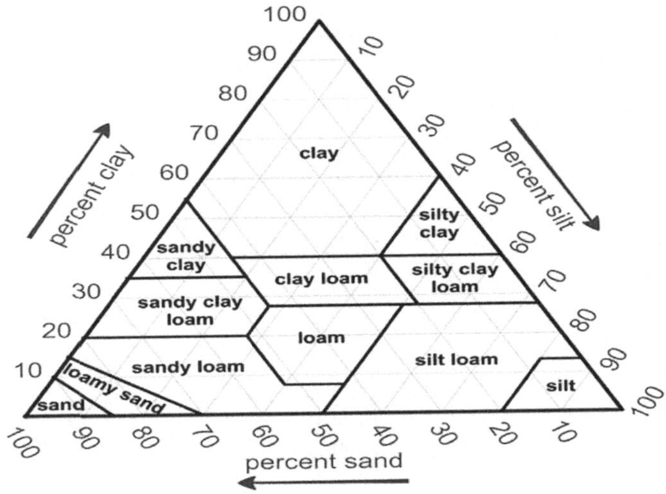

Figure 2: Soil Texture Triangle

Calculate the percent of sand, silt and clay using the temperature corrected hydrometer readings and the following equations:

- Oven dried soil (g) = (50 g x percent oven dried weight)/100
- Silt +clay (percent) =(Corrected 40 sec hydrometer reading x 100)/oven dried soil(g)
- Clay percent = (Corrected 2 h hydrometer reading x 100)/oven dried soil (g)
- Silt percent = percent (silt + clay) - percent clay
- Sand percent = 100 - percent (silt + clay).

To obtain the textural class of a soil sample use the soil texture triangle.

Properties of Soil

The properties of soil are determined by the composition of the soil, depending on different amounts of biotic and abiotic components. The combinations of these components determine the physical and chemical properties of soil.

Physical Properties

The physical properties of soil include the following:

i) Soil Texture

- Soil texture refers to the size of the soil particles that is dependent on the relative proportions of mineral components like sand, silt, and clay.

- ☆ Soil texture is further influenced by soil porosity, infiltration, and water retention capacity.
- ☆ The texture of soil differs with soil type; sandy soil feels gritty, silt feels smooth, and clay is sticky and mouldable.

ii) Soil Structure

- ☆ The textural components of soil, including sand, silt, and clay might result in aggregates as a result of their clumping. The aggregates further clump together to form peds.Information on the structure of soil gives information on the soil texture, matter content, and biological activity.
- ☆ Soil structure is influenced by physical processes that might be improved or destroyed by the choice of farming practices.

iii) Soil Density

The average soil particle density ranges from 2.60 to 2.75 g per cm^3, which usually remains unchanged for a given soil.

- ☆ The soil particle density is lower for soils with high organic matter content and higher for soil with higher mineral content.
- ☆ Soil particle density is different from soil bulk density which is always less than soil particle density.
- ☆ Soil density usually depends on the soil texture and structure and the composition of the soil.

iii) Soil Porosity

- ☆ Soil porosity is defined by the number of pores present within the soil.
- ☆ The porosity of soil is determined by the movement of air and water within the soil.
- ☆ Healthy soils usually have more number of pores between and within soil aggregates, whereas poor quality soils have few pores or cracks. Soil porosity is influenced by soil texture and structure. The pore size in soil affects the ability of plants and organisms to access water, oxygen, and other gases and minerals.

iv) Soil Consistency

- ☆ Soil consistency refers to the ability of the soil to stick to itself or other objects and to resist deformation and rupture.
- ☆ Three moisture conditions define soil consistency; air-dry, moist, and wet.
- ☆ The consistency of dry soil ranges from loose to hard, whereas that of wet soil ranges from non-sticky to sticky.

- Soil consistency is an important property that determines the ability of soil to support buildings and roads.

ii) Soil Colour

- Soil color is determined primarily by the organic composition of the soil.
- Soil color is one of the factors that help in the prediction of other soil characteristics within a soil profile. Observation of soil color is a qualitative means of measuring organic, iron oxide, and the clay contents of the soil.
- Besides, soil color is also influenced by the mineral content of the soil as the color might change as a result of oxidation of degradation.

Chapter 5

Pond: Designing and Construction

Ponds are large water bodies, typically of earthen construction and usually smaller than lakes. Throughout the world, earthen ponds are the most used unit for fish production. More than 40 percent of world's aquaculture production is carried out in ponds (Nash, 1988). Ponds are used both for fish and shell fish and also for seed production and raising the seed to marketable size.

Pond ecosystem is a natural ecosystem. It includes full algal photosynthesis. During the day time, the algae produce oxygen by photosynthesis, while during the night time they consume oxygen. Thus, there will be a daily fluctuation in the dissolved oxygen level in the pond, and special care must be taken to maintain the oxygen level. Similarly, the pH may fluctuate because photosynthesis fixes carbon dioxide and therefore, pH will increase during the day time, while at night, the algae release carbon dioxide and the pH will drop. The pond ecosystem will affect all nutrients since a nitrogen cycle will occur in the pond; Nitrification will transform total ammonia nitrogen to nitrite (NO_2) and further to nitrate (NO_3). If there are areas in the pond lacking oxygen, denitrification of nitrate to molecular nitrogen will also occur.

A major benefit of a pond is that it is possible to utilize the biological production, which includes prey that is food for the fish. In addition, there will also be decomposition of waste. Thus, water exchange rate is very less. The major disadvantages with production in ponds are the low production per unit volume, and the difficulties of maintaining control over the water quality and the actual fish biomass.

Types of Ponds

There are several ways to classify the ponds. They are based on shape, construction, drainage facility and water supply.

A. Pond Types Based on Shape

There are types of ponds based on shape. They are

- ☆ Rectangular
- ☆ Circular
- ☆ Oval
- ☆ Irregular

B. Pond Types Based on Construction

Based on construction, three types of ponds can be identified.

i) Dug-out or Excavated Pond

It is constructed by removing soil from an area to form a depression or hole and filling it with water to form a pond. Any pond shape can be formed by digging, but rectangular shape is most common because they are easier to form with excavation equipment. These ponds can be constructed on level terrain, a characteristic of many coastal structures. These ponds have several limitations.

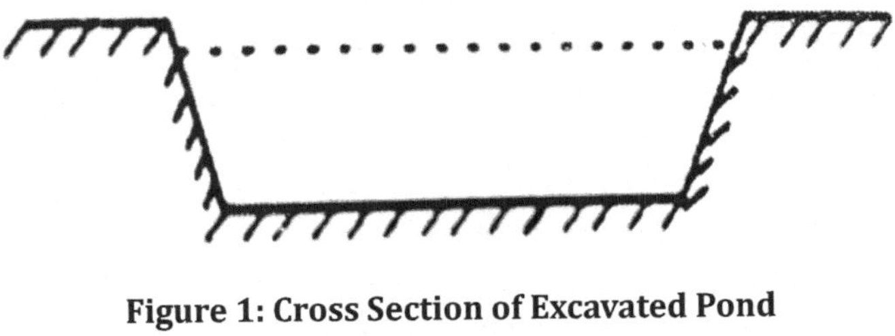

Figure 1: Cross Section of Excavated Pond

- ☆ Water removal is to be done by pumping.
- ☆ Unless dykes are constructed, flooding by high tides may occur, resulting in complete crop loss and/or introduction of unwanted species.
- ☆ Additional expense is to be incurred to remove settled faecal materials.
- ☆ Mechanical aeration is required to maintain dissolved oxygen levels.
- ☆ Harvesting is difficult.

ii) Embankment or Levee Pond

Embankment ponds are formed by building up a dam, dike and similar above ground structures to impound water. These are the most common ponds used for fish culture, primarily because they can be constructed in a great range of topographic conditions, especially flat land shape, and large areas can be used for this type of pond

There are two ways to establish such ponds.

- ☆ Those which can be at ground level, or,
- ☆ The levee can be above the ground level and the bottom below the ground level.

For the first type, it is necessary to bring the soil from outside to build the dyke; for the second type the excavated soil can be used to construct the levees which will reduce the cost of establishing the pond. It is important that the levee/ dyke is sufficiently wide to carry traffic, *i.e.* for feeding, maintenance or harvesting.

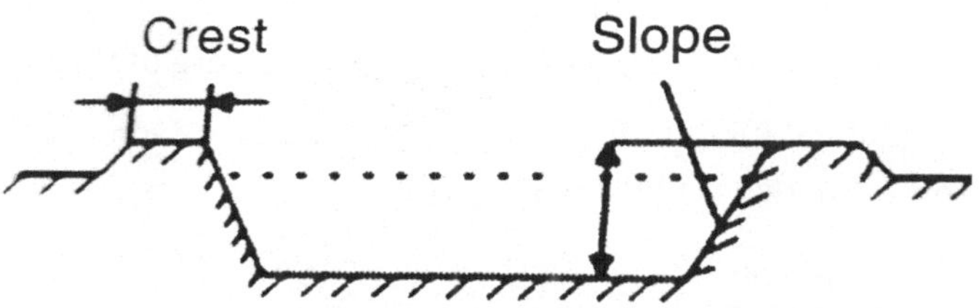

Figure 2: Cross Section of Levee Pond

During construction of this type of pond, it is possible to tap the water to a lower level to make easy drainage through gravity. Eventually a canal drainage system may also be excavated in conjunction with the pond area.

iii) Ravine Pond

These ponds are also called as watershed ponds. Watershed ponds utilize the terrain features mostly taking advantage of natural slope of hill and by constructing a dam on lower side. Since there are very few sites that satisfy the requirements for a watershed pond, this is not a very common pond type.

Pond types based on drainage facility: Based on drainage facility, the ponds are categorised into two types.

a. Drainable Pond

- ☆ These ponds have control over the water level.
- ☆ It is simpler to fertilize/feed and also to supply aeration.
- ☆ Cultivation practices can be run more intensively depending on amount and growth rate of fish.
- ☆ These ponds have effective harvesting process.

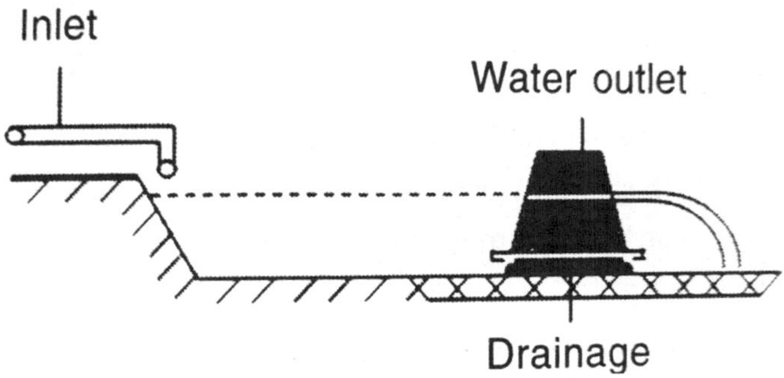

Figure 3: Drainable Pond

b. Non-drainable Pond

- Normally larger ponds up to several hectares are of this category
- Low establishment cost
- Culture practice is of extensive type

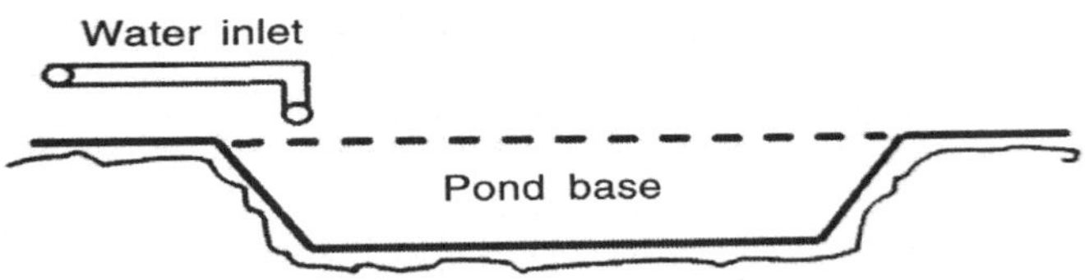

Figure 4: Non-Drainable Pond

Pond types based on water supply: There are three major types of ponds according to their water supply. They are:

a) **Spring water ponds:** In this case water is supplied from a spring in the bottom of the pond, or in close proximity to it, or supplied by groundwater. The groundwater used in ponds, lack dissolved oxygen content. Hence, provisions to increase oxygen content in water are highly necessary.

b) **Ponds supplied with rain water or by run-off water:** Ponds depending on rain water are called as 'sky ponds'. After being drained dry, these ponds are filled up more or less quickly according to its size and to the amount of rainfall.

c) **Ponds supplied by a water course:** These are classified into two types based on whether the entire water course or only a part crosses the ponds.

   i) **Barrage ponds:** Ponds supplied by springs and by rain water are known as dam or barrage ponds. There is no by-pass for evacuation of excess water. Ponds overflow successively one into other. It might be used in places where there is no danger of a rush of water such as ponds fed by springs.

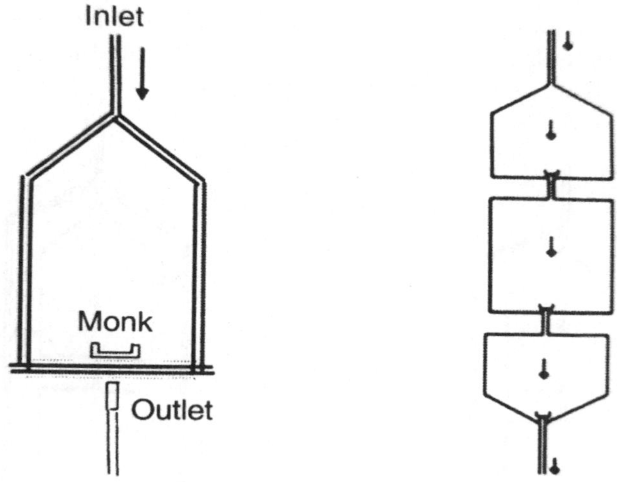

Figure 5: Barrage Ponds

ii) **Diversion ponds:** By-pass channels are used for the evacuation of excess water.

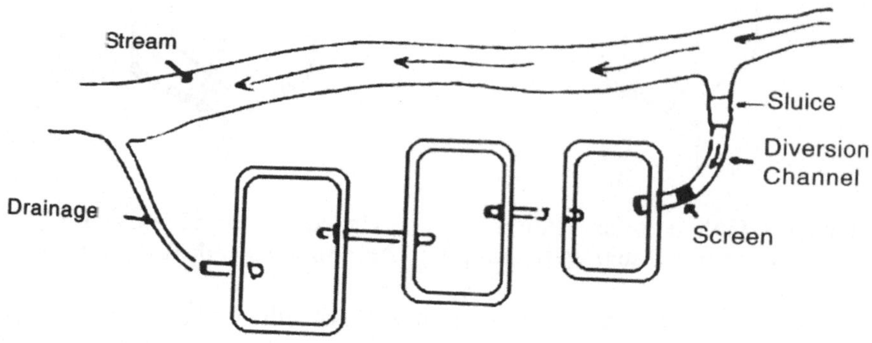

Figure 6: Diversion Ponds

These are also called as 'Rosary system'. Diversion ponds are subdivided into two types. They are a linked ponds in a row: A by-pass channel evacuates excess water. Ponds overflow successively one into the other. This must be adopted for valleys with rather steep longitudinal slopes or when the water flow is insufficient.

a) **No supply channel (Linked ponds without autonomous and individual water supply):** A single by-pass for evacuating excess water. Disposition does not permit individual and autonomous supply nor the isolation of ponds.

b) **A supply channel (Linked ponds with autonomous water supply):** There will be two by-passes; One for supply, the other for the evacuation of excess water and drainage water. This is preferable to the preceding method and permits the individual supply of each of the ponds. In addition, the overflow of water from one pond to other is possible.

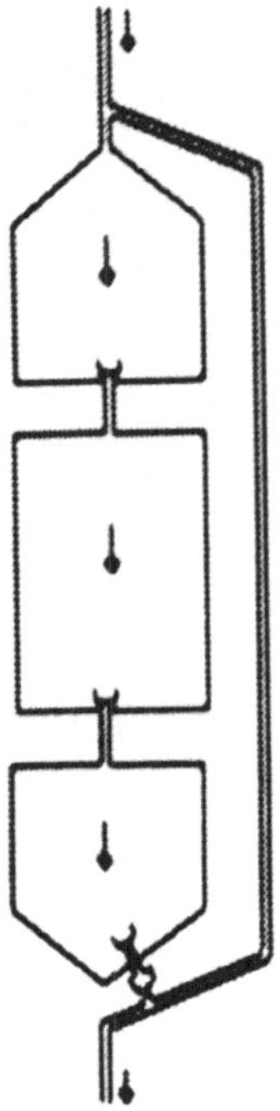

Figure 7: Linked Ponds without Autonomous and Individual Water Supply

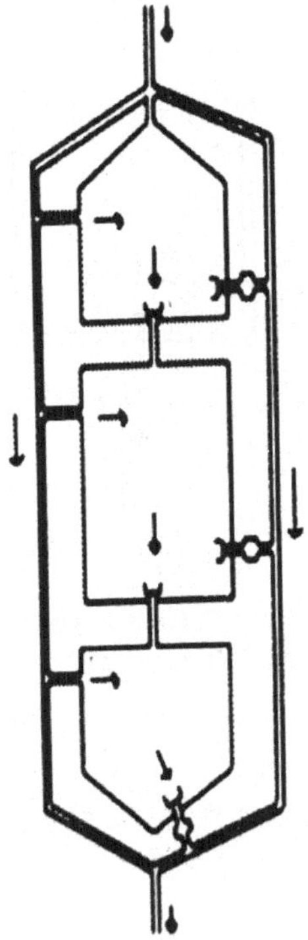

Figure 8: Linked Ponds with Autonomous and Water Supply

iii) Parallel ponds: In these ponds, there will be two by-passes; one for supply and the other for evacuation. Supply channel is for feeding the pond with water, and by-pass channel for the evacuation of excess water and for emptying the pond. Each pond has its own independent supply and evacuation. This type of ponds is suitable in valleys with gentle longitudinal slope and the disposition is ideal as long as flow is sufficient.

a) **One series of ponds:** These ponds can be built in relatively narrow valleys

b) **Two series of ponds:** These ponds can be built in wide valleys.

Planning of Fish Pond

Earthen fish ponds are the most used unit for fish production. The main purpose of an earthen pond for aquaculture farming is to create a restricted volume where fish or other aquatic organisms can be fed in a good water environment

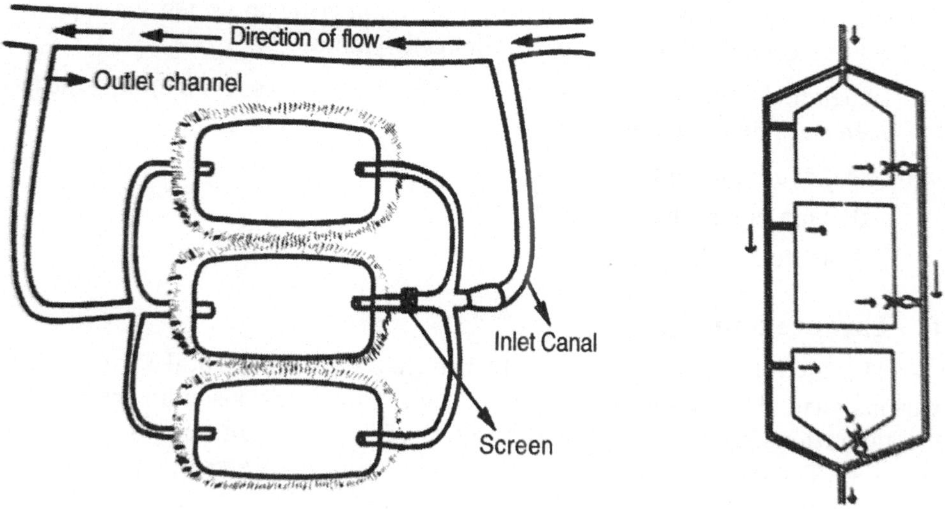

Figure 9: One Series Parallel Ponds

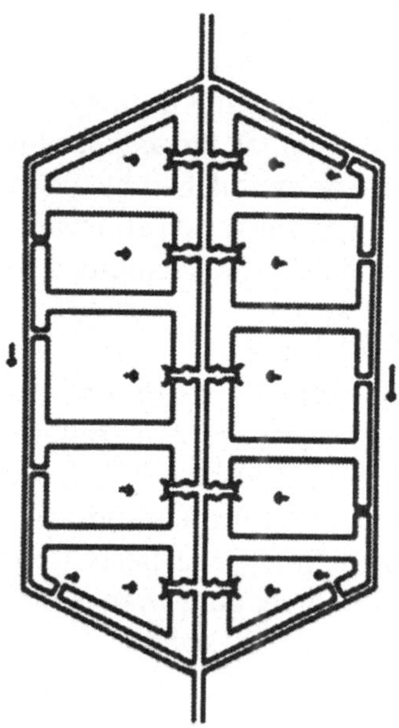

Figure 10: Two Series Parallel Ponds

and thus a natural ecosystem is being created inside. Different types of fish ponds may be used for fry production, fingerling production and on-growing fish ready for marketing. A fish farm may produce eggs, fry, juveniles or on-growing fish for market or, it can have a complete production system with all life stages from eggs to harvesting on the farm. In all these types, difference is normally the size and depth of the pond. Besides, the aquaculture systems range from extensive, through. Semi-intensive to intensive. Hence, layout of farm, design and construction of facilities require careful consideration of engineering and scientific aspects to make it functionally efficient and economically viable.

Depending upon the nature of work involved, the process of planning the farm has been classified into following steps:

- ☆ Layout planning
- ☆ Man power planning and
- ☆ Material planning.

i) Layout Planning

The layout planning deals with the distribution of total farm area for various farm elements depending upon the requirement based on the scientific considerations. The layout should be prepared per the land configuration and contours. The deep depressed area should be utilized for stocking pond, slightly higher land for rearing pond, moderately higher land for nursery pond and in higher contours, rest shed, store house *etc.* should be built. This will involve less quantity of earth work and good storage of water. For economic design, the earth available from excavation or digging should be equal to the earth required for raising the dyke. The layout of water supply and drainage net work should be prepared according to the contours, so that, as far as possible the water supply and the drainage arrangement can be accomplished by gravity. The other infrastructure facilities like roads, electricity, office, watch and ward, stores, marketing area, *etc.* should be appropriately located so as to facilitate operation and management.

ii) Man Power Planning

This deals with economical aspects of farm construction which includes the estimation of cost of the project and time required for the purpose. For this, calculation of total number of man days (skilled, unskilled labours,supervisors, technical and scientific personnel) for the project to complete should be decided before hand. Since it deals with financial aspects, knowledge on the locality,socio-economic conditions of the people, daily wages paid, time period required for various farm elements is very much desired.

iii) Material Planning

It deals with the procurement of construction, materials and equipments. Generally good quality with low cost investment should be the motto. Further, maintenance of the materials/stores purchased is another important aspect.The materials required need to be purchased before hand and stored in safe so that they will not hamper the work while in progress.

Shape of the Pond

The shape of the pond varies mainly on the topography of the area. Although various shapes of ponds such as circular, square, rectangular and oval are in use, rectangular and square ponds are in common use for commercial aqua farming systems. Square or round ponds may be cheaper in construction for their shorter length of dyke, but not preferable, because very wide ponds are difficult for harvesting. Besides, construction of a number of round ponds in an area will cause wastage of land in between.

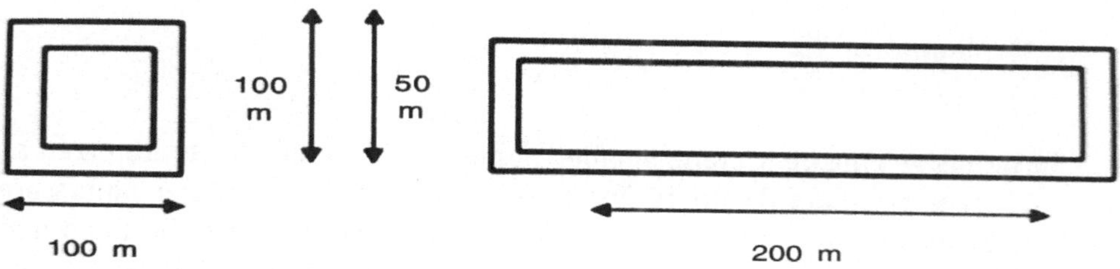

Figure 11: Length of the Dyke for Square and Rectangular Pond

Generally in freshwater tanks, rectangular ponds are the most preferred ones, as they facilitate operating the nets and maintaining regular slope of the pond bottom. For ideal rectangular ponds, the ratio of length and breadth should be 2:1. The width of a stocking pond (rectangular shape) may not normally exceed 50 m, so that relatively smaller net and limited number of men would be sufficient for harvesting.

In rectangular ponds, the larger side must be oriented parallel to the direction of prevailing wind for better aeration. But, at places where wind velocities are very high (more than 6.5 km/h), it is recommended to design the longer side vertical to the direction of prevailing wind to prevent soil erosion.

Size of the Pond

The size of individual ponds determines the overall ratio between the total area of the farm and the area covered by water. Pond size mainly depends upon the prevailing culture practices with production goals, expected income and marketing strategies. Pond size typically range as follows:

Table 1: Optimum Size for different Types of Ponds

Sl.No.	*Type of the Pond*	*Size (ha)*
1.	Nursery pond	0.02 to 0.06
2.	Rearing pond	0.06 to 0.1
3.	Stocking pond	0.2 to 2.0
4.	Marketing pond	0.2 to 0.4

A number of factors form the basis of deciding the size of ponds in aquafarms. These include

a) Quantity of available water

b) Land area
c) Technology to be followed (*e.g.* extensive, semi-intensive and intensive farming)
d) Production and income
e) Access to marketing
f) Manpower
g) Equipment, *etc.*

In intensive culture system, smaller ponds of 1.0 to 5.0ha are preferred as against larger ponds of 3.0 to 10.0ha in extensive systems. Smaller ponds are convenient to operate where as larger ponds take longer time to be filled and drained under a given set of conditions. Though convenient, the construction cost of smaller ponds is more per unit area to the additional cost of the embankments and water supply arrangements.

Advantages of Large Ponds

- ☆ Cost of construction in less per hectare of water spread area
- ☆ Area required is less per hectare water spread area
- ☆ Availability of oxygen is more in the water

Disadvantages of Large Ponds

- ☆ Filling or emptying of ponds takes long time
- ☆ High maintenance cost
- ☆ Difficulty in controlling diseases
- ☆ Harvesting is difficult

Advantages of Small Ponds

- ☆ Harvesting can be done easily and quickly
- ☆ Drain and refill can be done easily
- ☆ Treatment for disease is easy
- ☆ Erosion due to wind velocity is less

Under Indian condition, the distribution of nursery, rearing and stocking ponds should be 5 percent, 20 percent and 75 percent respectively. A size of pond ranging between 1.0 to 2.0 ha is considered good from management point of view.

Small ponds are easier to manage and fish production per unit area is greater. However, in smaller ponds, embankments and drainage ditches occupy considerable space, and thereby increases the cost of construction. In aquafarms with smaller ponds and enclosed by the periphery embankment, the total productive area of the farm will be reduced (*i.e.* the smaller the pond, the greater is the area occupied by the embankments and drainage ditches). Generally, the embankments and the drainage ditches constitute about 20-25 percent of the total productive area. Therefore, bigger ponds are to be designed which will reduce the area occupied by embankments. The size of pond considered good for management is between 1 to 2 ha.

The area of the storage pond is determined by the amount of culture species expected to be stored. A good pond can hold about 1 Kg/m^3 with no special arrangements and up to 15 Kg/m^3if continuous water supply and aeration is provided.

Types of Ponds

There are mainly three types of ponds, namely nursery ponds (N), rearing ponds (R) and stocking ponds (S). The ratio of water area between these ponds depends upon actual use of the aquafarm. For example, in case of fish seed farm, mainly nursery and rearing ponds are required with some nominal area for brood stock ponds. Whereas, in case of fish production farms where fully grown fishes are produced after procuring and stocking fingerlings, stocking ponds are required in large number with a few small ponds for multi-purpose use. For an independent fish farm, the sizes of nursery, rearing and stocking ponds will depend upon stocking density in individual fish ponds. There is no standard recommendation or formula for distributing the water spread area for nursery, rearing and the stocking ponds. However, in common practice, the ratio of water spread areas of N, R, and S ponds varies between 1:2:4 to 1:2:8 (N:R:S). Now-a-days, the multi-purpose ponds, wherein both rearing and stocking can be done are becoming popular.

Nursery Ponds

Generally 10 to 15 percent of the total water spread area is reserved for nursery ponds. 'N' ponds are constructed with cement concrete or brick masonry, and all the bundhs will be completely lined with either cement concrete or stone slabs, N-ponds are smaller in size varying from 0.05 to 0.1 ha for fish ponds and upto 0.5 ha for shrimp ponds. Depth of nursery fish ponds vary in the range of 0.8 to 1.0 m.

Therefore, better aeration and water exchange are desired in nursery ponds. A size of 4.5m x 9.0 m to 10.0 m is generally recommended for nursery ponds. Nursery ponds can also be used as grow out ponds during off seasons. Nursery ponds are used for nursing of fry to juvenile stage. Clean and sufficiently oxygenated water is used for better survival of seed. After rearing the seed for 2-3 weeks (in case of shrimps), the juveniles are collected and stocked in the grow-out ponds.

Stocking Ponds

Stocking ponds occupy the major portion of the productive area/water spread area. About 60-65 percent of the total productive area is provided for stocking ponds. Stocking ponds are bigger in size (about 1 to 2 ha.) and deeper than other two types of ponds. There is no standard size for stocking ponds; however, the width of stocking ponds is less than 40 m in most of the fish farms. In case of fish ponds where harvesting is done by complete drainage (de-watering) of the ponds, no such restriction is imposed on the width of the pond. Stocking ponds may be of either rectangular or square shape. The length and breadth ratio of rectangular ponds varies from 1.5:12021 (LB), depending upon the stocking density, direction of predominant winds, managerial capacity and marketing potentials. The size of stocking ponds mainly depends upon the stocking density. A pond can hold fishes from 1kg/m^2 to 15 kg/m^2 depending upon the type of culture (*i.e.* culture technology).

Depth of stocking ponds mainly depends the quantity or volume of water to be retained in the ponds, type of culture (*i.e.* extensive, semi-intensive and intensive), mode of culture (mono culture or poly culture), and also the natural configuration of ground surface (*i.e.* topography).

For fishes, 1.2 to 1.5 m depth is ideal; whereas, for shrimps 0.8 to 1.0 m is sufficient in extensive system, and 1.0 to 1.2 m is desirable for semi-intensive system (as stocking density is very high).

Stocking ponds may be constructed with uniform bed or with central platform and periphery trench. In cases where the excavation of 20-30 cm for the ponds meets the requirements of earth for bundhs and maintenance of desired water level in the ponds, it would be better to construct the ponds with uniform bed. In any case, an excavation more than 50 cm on the ground level is uneconomical.

- For stocking ponds with uniform bed, a longitudinal slope of 1:1000 is necessary towards an outlet for quick discharge of pond water. A small harvesting pit is constructed for collecting the fishes nearby any outlet in the ponds where harvesting is done by total de-watering of the pond.
- Ponds with central platform and periphery trenches would be suitable where ground level is higher than the desired water level of the pond or if tidal levels are not convenient to maintain at least 90-100 cm water level in the ponds.
- In such cases, it is desirable to leave the central platform untouched and form the bundhs with the excavated earth of periphery trenches only
- For economical ponds, excavation of trench should be between 50 to 60 cm, and breadth can be adjusted accordingly to obtain the required quantity of earth for formation of the bundhs.
- Each stocking pond should be designed with at least one inlet and one outlet, preferably placed in a zigzag fashion for thorough mixing of the feed.
- In case of semi-intensive farms two inlets and two outlets may also be provided for ponds bigger than 2 ha. in size for quick discharge of water when water quality gets deteriorated
- Corners of the ponds are made round and smooth for better circulation of water and to prevent the soil erosion in the corners.

Design of Pond

- Size
- Shape
- Depth
- Bottom of the slope
- Construction of pond
- Marking
- Excavation of pond

i) Size

The shape, size and orientation of a pond are totally depend upon topography, property lines, construction, and operating cost, available resources and management ease. The size of individual ponds determines the overall ratio between the total area of the farm and the area covered by water.

The farm should be affordable and economical to construct and operate. Pond size should be compatible with production goals, expected income, and marketing strategies. Sizes vary depending upon their use (*i.e.*, fish production, fingerling production, brooder pond, spawning, *etc.*). Pond size typically range as follows:

- Nursery pond: 0.05 to 0.20 ha
- Brood stock or Grow-out ponds.: 0.25 to 10.00 ha
- Spawning ponds: 0.01 to 0.05 ha
- Intensive culture ponds: 1.00 to 5.00 ha

Small ponds are easier to manage and fish production per unit area is greater. However, in smaller ponds, embankments and drainage ditches occupy considerable space, and thereby increases the cost of construction. In aqua farms with smaller pond and enclosed by the periphery embankment, the total productive area of the farm will be reduced (*i.e.* the smaller the pond, the greater is the area occupied by the embankments and drainage ditches). Generally, the embankments and the drainage ditches constitute about 20-25 percent of the total productive area. Therefore, bigger pond are to be designed which will reduce the area occupied by embankments. The size of pond considered good for management is between 1 to 2 ha.

The area of the storage pond is determined by the amount of culture species expected to be stored. A good pond can hold about 1 Kg/m^3 with no special arrangements and up to 15 Kg/m^3 if continuous water supply and aeration is provided.

ii) Shape

The main factors affecting the desired shape of the ponds are:

a) The ratio between the length of dike and the area covered by water. This affects the cost of construction of aquafarm
b) The topography of the area
c) The method of harvesting

The method of harvesting will affect and sometimes determine the shape of the ponds. If harvesting is done by netting, pond width should not be greater than 100 meter. Greater widths demand expensive changes in length and strength of fishing nets, increased manpower, and towing equipment. This may negate the cost saving in a wider pond. Although there are many options (circular, square, rectangular and oval) in designing the shape of ponds, from practical utility and management point of view, rectangular ponds and square ponds are in common use for commercial aqua farming systems.

For fish culture where the harvesting is done by nets, rectangular ponds are preferred since they can be constructed adjacent to one another making water supply and management practices easier. For ideal rectangular ponds, the ratio of length and breadth should be 1.5:1 to 2:1 (L:B). In rectangular ponds, the longer side must be oriented parallel to the direction of prevailing predominant wind for better aeration. But, at the places where the wind velocities are very high, soil erosion of the dikes takes place. Therefore, in such places, it is recommended to design the longer side perpendicular to the direction of prevailing wind. This will help in mixing of pond waters as wind travels across the pond. Rectangular tanks and troughs are most often used for rearing of fry and fingerlings in hatcheries. In tanks that drain at one end, the water enters from a different end, much like a raceway. Flow patterns often form dead areas like tank corners. Fish may become stressed, if not die, if metabolic wastes accumulate to dangerous levels. Therefore, the corners of ponds must be made round and smooth.

Square ponds are most economical of the geometrical figures from the point of view of construction. In a square pond, the ratio of water area to the length of the embankment is highest; hence the cost of construction is lowest. When there are no topographical limitations and harvesting is done by total draining, square ponds are recommended. But, on steep grounds, draining of square ponds is problematic and time consuming also. For shrimps and prawn culture, it is recommended to construct square ponds as they provide better aeration uniformly throughout the year with the wind blowing from any direction resulting in lesser erosion and therefore planning would be much easier.

Round tanks are commonly used for nursery. Round concrete tanks are used for intensive culture systems. These tanks have either flat or sloping (conical) bottom. Water circulation in the round tanks is excellent and is very efficient at removing the wastes. Round tanks have central stand pipes that are fitted with outer pipe openings near the tank bottom facilitate waste removal. Screens or grader bars that can be rotated around the central stand pipe facilitate the removal of fish by crowding them in to a small area where they are netted. Harvesting is done by netting, pond width should not be greater than 100 meter. Greater widths demand expensive changes in length and strength of fishing nets, increased manpower, and towing equipment. This may negate the cost saving in a wider pond. Although there are many options (circular, square, rectangular and oval) in designing the shape of ponds, from practical utility and management point of view, rectangular ponds and square ponds are in common use for commercial aqua farming systems.

iii) Depth

Pond depth mainly depends upon the species to be cultured, topography of the area, and the climatic conditions. Deeper ponds are difficult to manage. Shallow ponds get too hot and support the growth of rooted aquatic plants and decrease the effective area of the pond and also impede seining. Otherwise for fish culture 86 to 90cm depth is sufficient. When the water supply is seasonal, it is necessary to increase the water volume by increasing the depth of the pond so as to accumulate enough water during the wet period to suffice for the entire growing season. Yield in the deeper ponds are high due to better temperature regime in the ponds and dilution of metabolites excreted by the fish. Deep ponds have some disadvantages also.

For example, the nets have to be deeper. Following table gives the depth requirement of a few species.

Table 2: Depth Requirements of a Few Species

Sl.No.	*Species to be Cultured*	*Pond Depth in meters*
1.	Cat fish	0.9 to 1.5
2.	Major carps	1.5 to 2.0
3.	Shrimps	0.75 to 0.9
4.	Craw fish	0.5 to 0.6
5.	Freshwater prawns	0.75 to 0.9

Storage depth is the amount of storage volume required to retain any additional water due to storms and floods to prevent damage to dikes or complete failure of the dam. Storage height is the difference in elevation from the crest of the mechanical spillway to the emergency spillway.

The sizes of individual ponds determine the overall ratio between the total area of the farm and the area covered by the water.

iv) Pond Bottom

The pond bottom should be smooth and slope even in order to drain out the water completely and also to concentrate fish in deeper part of the pond. A minimum bottom slope of 0.1 to 0.2 percent is sufficient. An uneven bottom develops depressions where fish can take shelter making fish catching difficult. It is important that pond bottoms are dried out and renovated after each harvest. For complete drainage, the deepest part of the pond near the monk (*i.e* outlet gate) should be deeper than pond bottom level by as much as possible. This deepest part of the pond (the harvesting sump) should not be greater than 0.1 ha, and in small ponds, it can be much smaller for ease in harvesting. The most important characteristic of harvesting sump is its size and depth. It should be able to contain all the fish in pond just prior to harvesting. The position of harvesting pit is also important. It should be protected from siltation. It should have a supply of freshwater and aeration for the period the fish are concentrated in it.

The pond bottom slope requirement puts a practical limit on the pond size. Steep slope even in small ponds and gentle slope even in large ponds can result in a wide range of water depths throughout the pond. This can lead to undesirable thermal and chemical stratification in deeper water areas and undesirable plant growth on the bottom of the shallow water areas.

v) Process of Construction

The first stage in the construction process involves general planning. For this purpose, a topographical map of scale 1:2000 to 1:2500 is required. General features of the farm such as division of the entire farm area into various ponds and their types (rearing or auxiliary ponds), the number of ponds, size of the ponds, the orientation of the ponds, the width of the embankments, and the drainage and water supply systems are worked out on this map.

The next stage involves detailed planning of each pond. For this purpose, a topographic work plan of larger scale is required. This will show where and how much to excavate, and where the earth should be moved. For this purpose a survey has to be made. The pond area is divided into 20 x 20 m squares by pegging the ground at these intervals. The peg line should be parallel to the direction of the longest embankment of the pond. If the ground surface between any two pegs is not regular and changes drastically by reason of a ditch or mound, a peg is added at the place of change. These pegs form a network of coordinates in which each peg is marked by a letter and a number. The ground level at each of the pegs and the levels of the peg heads are measured, and the elevations are noted on the work plan of the pond.

Next, the amount of earth required for construction of embankments is calculated from the general plan. This is done by determining the amount of earth required for 1 m of embankment (this is equal to the volume of 1 m trapezoidal cross-section of the embankment having the planned dimensions) and multiplying by the total length of embankments. The packing coefficient of the soil should be taken into account. This coefficient depends on the type of soil and the kind of construction equipment used. The heavier the equipment, the greater is the packing coefficient, usually 20-25 percent. This means that when 1 m' of earth is required for the packed embankment, 1.2-1.5 m^2 should be actually provided.

Generally, about one-third to one-quarter of total pond depth is obtained by excavation. The rest is obtained by depositing earth on the area earmarked for embankments and building them up. The amount of earth which can be excavated is calculated and compared to the amount required (again taking into account the packing coefficient). If these amounts do not balance, tentative excavation depth is changed; accordingly and the amount of earth obtained is recalculated. According to Pruginin and Ben-Ari (1959), 2500-4000 m^2 of earth is required per hectare for construction of pond.

After excavation depth is calculated, the details are transferred to the working plan. The depth of the planned bottom from the head of the pegs is marked. The amount of earth dug out from each square between pegs is calculated, and the place on the embankment where the earth should be deposited is designated. This should be as near as possible to the place of excavation. The contours of the embankment's base are pegged in the field and construction is started according to the working plan

At the onset of construction, vegetation and roots should be removed. These should not be put into the embankment since organic material will rot and leave behind hollows which become weak points in the embankments and are the first to erode. Generally, top 10 cm soil layer is not used for construction as this soil will form weak bondage between the dike base and ground level. However, since this is fertile and productive, it can be placed on the pond bottom to increase pond productivity

The drainage pipeline is laid before starting construction of the embankment. The location and length of the pipeline are designed on the working plan. The pipeline should be covered by earth, which is then carefully and methodically compacted by hand tools to prevent leakage along the line. Such leakage may lead to bursting of the embankment in the area of the pipeline. An additional precaution is the construction of an anti-seep collar made of concrete or metal in the embankment around the pipeline and perpendicular to it extending to both sides. The pipeline should be covered with enough earth to protect it against damage from the heavy machinery passing over it.

During the construction work, special attention should be given to the shape of the embankment to make sure that its width at all heights corresponds to the plan. If the embankment is constructed too narrow, this cannot be corrected later.

vi) Land Marking

After the land is cleared off then the land is to be marked for the construction of various farm elements as per the layout planning. This operation involves laying out the features of ponds on the ground in order to mark out the area from where the earth will have to be cut or removed, and also where earth will have to be embanked. Initially lines are drawn according to the layout followed by pegging and fixing stakes or posts. Strings are stretched between the top of pegs and posts to mark the complete profile of the dyke with its correct height, width and slope. Dry white lime powder is used to show the positions of various farm elements like dykes, channels and ponds which are to be constructed. It is called as staking.

vii) Excavation of Soil

The excavation of ponds may be done either engaging manual labour, when speed of construction is under consideration bulldozer or scoop may be utilized. The excavation of the pond should be as far as possible preceded in layers. The top ground soil should be stocked in one place and it should not be used for any construction. Few dead men should be left on the pond bed for measuring volume of earth work. Which should be removed and pond bed made even immediately after the measurements have been recorded in the measurement book.

Selection, Design and Construction of Dikes

Embankment (Dike)

An embankment is a basic structure involved in the design a construction of aquafarms. Foundation soil of dikes must be able to support the weight of dike and vehicular load coming on the dike. Swampy, muddy or plastic soils should be avoided and dike must be constructed on the underlying consolidated soil material. Highly organic soils like peat and muck should not be used as foundation material as they decompose and cause the dam to settle and eventually fail altogether. Foundation material should have low permeability to prevent excessive seepage. Embankments are generally constructed using the earth available at the site. There are mainly four types of embankments:

i) Periphery Embankment
ii) Internal Embankment
iii) Compact dyke
iv) Partition or Secondary Dyke

i) Periphery Embankment or Main or Peripheral Dyke

Main dyke encloses the entire farm. It forms the boundary of the entire pond system. The purpose of this dyke is to retain water for culture operation as well as to prevent overtopping due to the flood, spring tides and storm and prevent failure due to slips and seepage. Periphery dike, which is also called as main embankment or perimeter embankment, protects the entire fish farm from floods, storm surges, and tides. The design of a periphery dike mainly depends upon the prevailing site conditions.

ii) Internal Embankment

Dikes constructed around the ponds are generally called as internal dikes. They are also known as secondary dikes. The height of internal embankment mainly depends upon the designed water level in the ponds. Necessary berms should be constructed in case the quality of soil is poor.

iii) Compact Dyke

It is always preferable to build a compact dyke. The dyke should be built in layers of 25 to 35cm in thickness, stretching across the whole section. Water should be sprinkled over the previous layers. Each layer should be rammed well until clods are flattened to have a good adhesion between successive layers.

iv) Partition or Secondary Dyke

The partition or secondary dyke is constructed in between the ponds limiting the size of each pond. These are generally smaller than the peripheral dyke. The height of partition dyke is calculated as;

$$H_p = H + H_w + H_f + H_s$$

where,

H_p = Height of the partition dyke (m)

H = Maximum desired depth of water in the pond (m).

H_w = Dyke height required for wave action (m)

H_f = Dyke height required for optimum free board (m)

H_s = Dyke height required for settlement (m)

The secondary dike is smaller than the main dike. It is constructed when the whole area is already enclosed by the main dike. The purpose of this dike is to divide the fishpond into rearing ponds, catching or head pond and supply canal.

The optimum free board of the partition dyke should vary between 0.3 to 0.5 m. For ponds up to 200m long, a minimum free board of 0.3 m should be provided where as for ponds between 200 m to 400 m long, minimum free board should be 0.5 m.

- ☆ Allowance for settlement is calculated as the percentage of partition dyke height. Average settlement allowance is between 10 to 25 percent of dyke height. Minimum allowance being 5 percent.
- ☆ The top width of 1.0 to 2.0 m is suggested as the movement is restricted to men and wheel barrows.
- ☆ The side slopes for inner dykes should be designed on the same consideration as that of peripheral dyke. The side slope is usually 1:1 to 1.5:1. Side berms may be provided if there is excess soil in order to reduce the cost of hauling.
- ☆ Puddle trench in dykes is provided whenever necessary.
- ☆ The computation of base width is done following the same method as in peripheral dyke.

Factors to be Considered in Designing a Dike

- ☆ High Flood Level (HFL) of the locality
- ☆ Type of soil
- ☆ Vehicular load expected on the top of embankment

- Slope of the embankment
- Designed water level of the ponds
- Purpose of construction
- Free board
- Site conditions or topographical features.

Important Design Parameters of a Dike

i) Free Board (FB)

Free board is provided as a safety factor of safety to pre- overtopping of dike. Free board can be defined as the vertical distance between the crest (top of dike) after settlement and the surface water level in the pond at its designed depth.

- Free board for main or periphery dike is normally 0.6 m
- Free board for secondary and tertiary dikes is 0.3 m

Allowance for wave action (H_{wa})

$$H_{wa} = 0.014(F)^{1/2} \text{ prime}$$

where,

F = Fetch of the wave

ii) Slipping of Dyke

The dyke of a pond is subjected to two types of forces.

They are:

i. Lateral thrust on the dyke and
ii. Surface wave due to wind

iii) Shear of the Dyke

It is also called as hydraulic gradient or, line of saturation or, line of percolation or, phreatic line. The hydraulic gradient of the dyke helps in stability of the dyke. The dyke shoul constructed such that, the hydraulic gradient should pass through the base of the dyke (AC) and there should be a clear cover of at least 30 cm between the line of saturation and the toe of the dyke (DO).

If the line of saturation passes beyond the dyke body (AB), then water will split out the dyke, or water will flow out. Thus, the dyke will be split out into two parts and the dyke may collapse.

Darcy's law states that, the rate of flow or discharge per unit time is proportional to hydraulic gradient.

$$V = K \times I$$

where,

V = Velocity of water flow

K = Coefficient of permeability

I = Hydraulic gradient

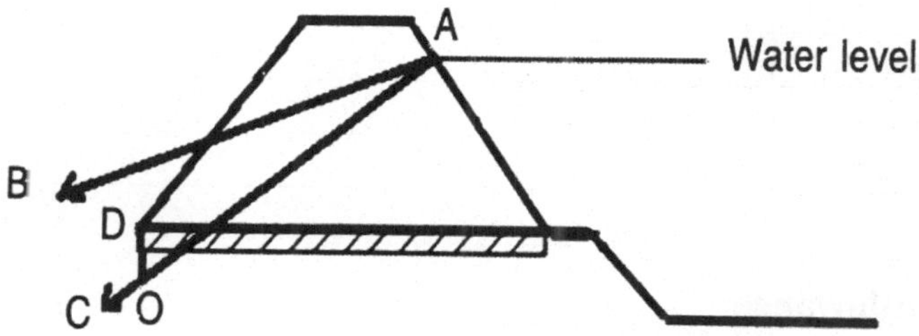

Figure 12: Desirable Hydraulic Gradient (AC)

Coefficient of permeability is defined as the average velocity of flow that will occur through the total cross sectional area of soil under unit hydraulic gradient. It is usually expressed as cm/sec or, m/day. The hydraulic gradients for different soil types are as follows.

Table 3: Recommended Hydraulic Gradient for different Soil Types

Sl.No.	*Types of Soil*	*Recommended Hydraulic Gradient*
1	Good clay good	1:3
2	Good compacted soil	1:4
3	Good compacted soil	1:5
4	Fine silt	1:6

iv) Slope of the Dyke

The soil also determines the slope of the dyke. A typical dyke is built with an outside slope of 1:1 and an inside slope of 1:1.5 to 2.0. A slope of 1:2 means that for every change in length of 2 m there is a change of I m in height.

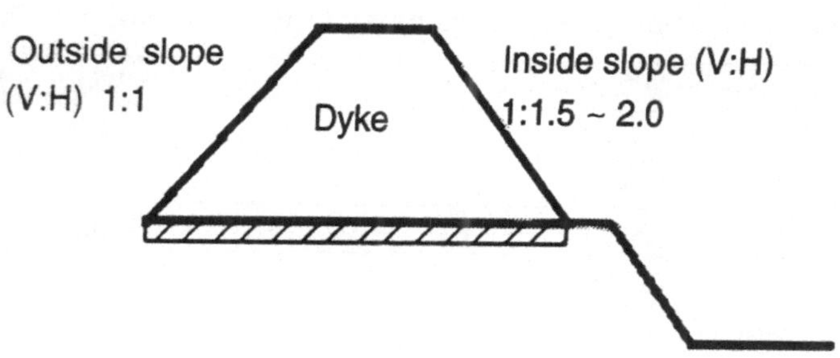

Figure 13: Desirable Slope of the Dyke

Table 4: Recommended Side Slopes for Earthen Dykes

Sl.No.	Type of Soil	Recommended Side Slope (Vertical : Horizontal)
1.	Good clay and silty clay	1:1.5
2.	Sandy clay	1:2
3.	Sandy loam	1:2.5
4.	Loose and wet earth	1:3

Settlement Allowance

It is the rate of settlement of the dyke (in height) in the first year of construction. In the subsequent years, it is negligible. A settlement allowance of 10 to 15 percent should be provided. The settlement allowance of the dyke has two parts.

$$h_s = h_w + h_f$$

where,

h_s = Settlement allowance

h_w = Allowance for wave action

F = Fetch

h_f = Allowance for freeboard

Top Width (B)

Top width (crest) of a dike is mainly depending upon the type of vehicular load expected on the top of the dike. The height of dike and its purpose are also important in designing the crest.

Cars, Jeeps *etc.*: 3-4 m

Tractors, Dumpers, *etc.*: 3.5-4.5 m

Trucks, Buses *etc.*: 4-5 m

Table 5: Soil Conservation Service Recommendations for Minimum Width of Dike

Sl.No.	Height of Dike (m)	Top Width (m)
1	Under 3.0	2.4
2	3 to 4.5	3
3	4.5 to 6.0	3.7
4	6.0 to 7.5	4.3

Height of Dyke

Total height of the dyke is determined using the formula:

$$H = (h + h + h_3 + h_4)\ m$$

where,

H = Safe height of the dyke (m)

h = Depth of water standing on the slope of dyke

h = Free board

h_3 = Settlement allowance

h_4 = Foundation/Seat of the dyke

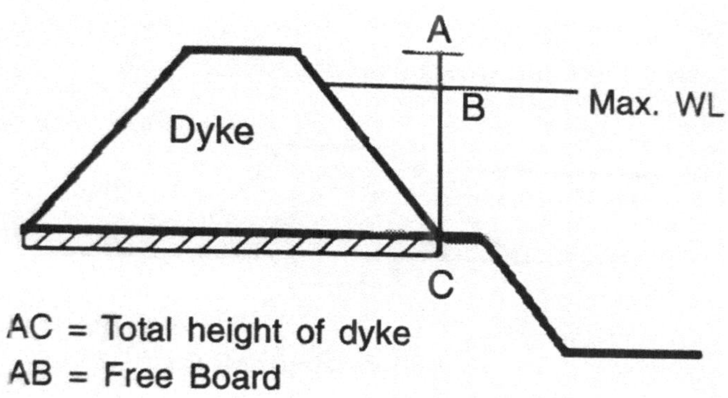

Figure 14: Total Height of the Dyke

Berm

It is the flat portion of the land between heel edge of the dyke (A) and edge of the excavation of main pond (B). Berm is essential for larger ponds. Desirable berm width in a large sized pond varies from 0.6 to 1.0 m

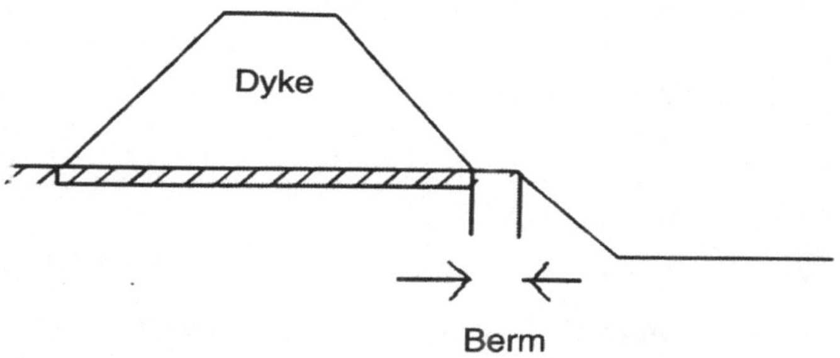

Figure 15: Berm of the Pond

Advantages of Berm

(*i*) During dragging the net, the fishermen will stand by the berm

(*ii*) This strengthens the pond's slope. If it is a continuous direct slope, it is not advantageous

(*iii*) The berm arrests the flow of soil to pond bottom

Construction of Dykes

The construction of dykes depends upon the kind of soil in which the pond is built. The soil, to be used for dyke construction, should be free from any organic should also be thoroughly cleared from the base of the dyke to allow binding of matter (roots *etc.*) as the rotting of the latter will weaken the dyke. The humus the dyke to the base properly and avoid seepage. The earth from salt affected areas should be avoided; sand and peatty earth also should not be used for construction of a dyke. Earth which is a mixture of sand, silt and clay having minimum 20 to 30 percent clay is found to be best for dyke construction. Earth obtained from pond digging is generally used for dyke construction.

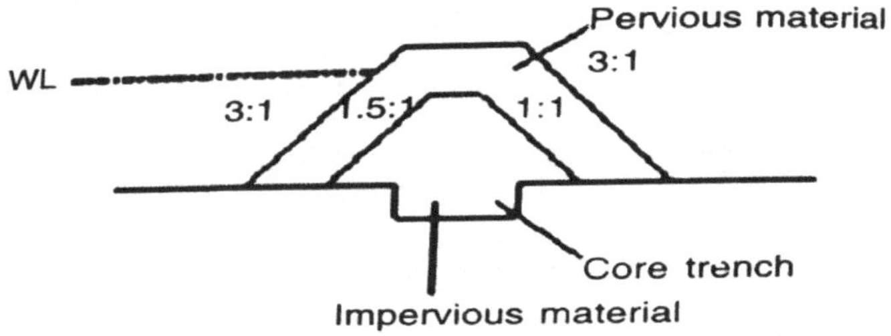

Figure 16: Core Trench Built on the Dyke of a Fish

The finished height of the dyke should be about 30 cm above water for small ponds and 50 cm above water for large ponds. The width of the dyke/wall at the top should be about equal to its height.

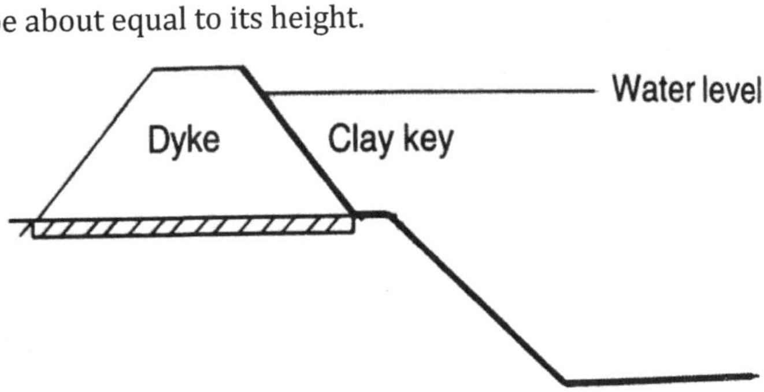

Figure 17: Clay Key to Prevent Seepage

The soil erosion of the rear side of the dyke can be checked by either stone pitching, or cement pitching, or glass turfing. This protection is to be done before start of the monsoon. Further, liquid asphalt is to be sprayed on it which has cement property.

Trees should not be allowed to grow at the dykes as the tree roots crack and destroy the wall. The desirable shape of the dyke is of trapezoidal.

The base width or bottom width of the dyke depends upon the depth of water, top width, side slopes and type of soil. The base should be sufficiently wide so that the seepage line should not appear above the toe on the downstream side of the dyke. The slope of the seepage line mainly depends on type of soil.

Table 6: Desirable Slope of Seepage Line for different Types of Soil

Type of Soil	*Slope of Seepage Line (Horizontal to Vertical)*
Clay soil	3:1
Sandy loam soil	5:1
Sandy soil	6:1

The side slopes of the dyke should be decided based on the soil properties and the velocity of water that may flow alongside the dyke during floods. The outer and inner slopes of the dyke depend on different soil texture. Unstable soils may require 4:1 or flatter slope.

Table 7: Desirable Side Slopes of the Dyke for different Types of Soil

Type of Soil	*Side Slope (horizontal to vertical)*
Clayey soil	1:1
Loamy soil	1.5:1
Sandy soil	2:1

Dyke Protection

- ☆ Pond dykes should be constructed with soil with optimum moisture content and in layers of not more than 30 cm thick. Each layer should be thoroughly compacted before the next layer is added. This helps for better compaction of dyke and to reduce the settlement.
- ☆ Compact dykes are preferred for brackish water farm. Mangrove plantation may be made as an additional protection of dykes which are exposed to strong wave action. The portion of dyke susceptible to wave action may be lined through turfing or sodding which is the most inexpensive way of dyke protection.
- ☆ Floating wave breakers maybe installed in longer ponds.
- ☆ Slope of the dikes should be lined with proper methods to prevent soil erosion.
- ☆ Stone pitching, brick tiling, concrete slabs, lime concrete mixtures, polymer based Chemicals *etc.* are some of the commonly used lining materials.

Chapter 6

Water Supply System

Water distribution system in an aquaculture farm involves the following:

1. Water distribution systems
2. Pipelines
3. Water control structure
4. Water budget equation
5. Pond drainage system

1. Water Distribution Systems

Canals

Open canals are suitable means of conveyance of water from adjoining source like river, steam, *etc.*, to fish ponds.

Types of Canals

Different types of open canals are used to transport water to fish farms. They are divided into four types in fish farms depending upon their purpose or use.

i. Feeder Canals

These are the canals used to supply water from the main water intake to the fish ponds. In a large farm with several diversion pond units there is usually a main feeder canal branching into secondary and even tertiary feeder canals.

ii. Diversion Canals

These canals are used to divert the required amount of water to fish ponds or to divert excess amount of water from (ponds) source.

iii. Drainage Canals

These canals are used to remove the unwanted or polluted water from fish ponds.

iv. Protection Canals

To divert run off water away from fish ponds or canals which carrying the water to ponds.

Construction of Feeder Canal

The most common type of water conveyance system is the open canal. In the open canal, water flows with a free water surface. Main open feeder channel supplies water to all the ponds. It starts from the main sluice gate or pumping station and generally traverses the central portion of the farm. In big farm, the main feeder channel could feed number of secondary channels also before water is conveyed to the ponds. Aquacultural farms usually have prismatic canals of rectangular, circular or trapezoidal shape. Most of the unlined earthen canals will have trapezoidal shape for stability.

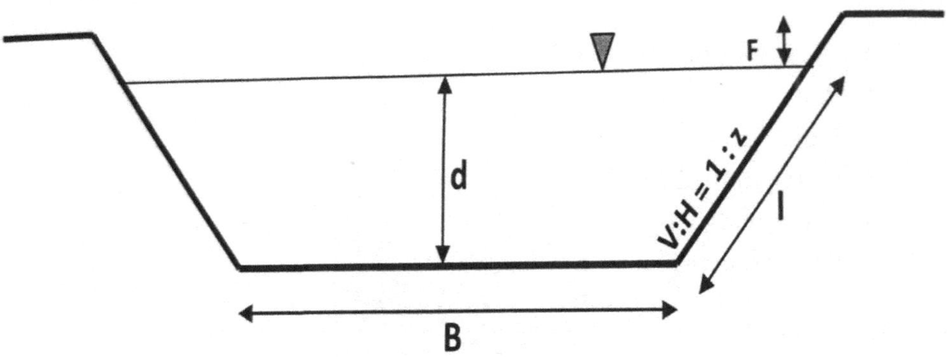

D: Depth of channel; B: Bottom width of channel; D: Depth of flow of water; Z: Side slope of channel; F: free bS: Bed slope of the channel; I: Wetted side of the channel; F: Free board.

Figure 1: Cross Sectional View of an Open Channel

The following geometric elements are defined:

i) Area of Cross Section (A)

It refers to the area of the wetted section of the open channel. In case of trapezoidal Section:

$$A = Bd + zd^2$$

ii) Wetted Perimeter (P)

It is the sum of the bottom and lengths of sides which are in contact with water.

$$P = B + 2l$$

iii) Hydraulic Radius (R)

It is the ratio between the cross-sectional area of flow and the wetted perimeter.

$$R = A/P$$

iv) Hydraulic Slope (S)

It is defined as the ratio of the drop head between channel sectors located at a distance.

v) Free Board (F)

It is the vertical distance between the top of the channel wall and maximum designed water level in the channel. The free board (F) is required as a safety measure against overtopping due to wave action or other unforeseen reasons.

2. Pipelines

The layout of the pipe lines should be as for as possible straight line. There should not be many bends in the pipe lines, which greatly reduces the discharge due to internal turbulence in the pipes. The metal pipes need to be protected with a coating of anti corrosive paint. Pipe lines passing through dams and marshy lands to be protected by canvas soaked in bituminous materials. The pipeline should be rest on at least 10 cm thick cement concrete bed with proper side packing on soft soil to safe guard against settlement. The diameter of the pipe should be large enough to discharge the required volume of water in specific time. The mouth of the supply inlet should be covered to prevent the entry of trash fish or predatory.

Types of Pipe Used for Water Supply in Fish Ponds

Following are the different types of pipes normally used in aqua farm.

- Plastic pipe
- R.C.C. pipe
- Plain cement concrete pipe
- Asbestos cement pipe
- G.I. pipe (Galvanized Iron)
- Cast Iron pipe
- P.V.C. pipes
- Copper pipes
- Wooden pipes

3. Water Control Structure

Definition

Gates are part of water distribution system in an aqua cultural farm and are installed to control the water flow. Following are the basic requirements of a gate.

- It should be of sufficient capacity to handle the required flow of water during pond draining and filling.
- It must be properly located and carefully constructed taking into the account the pond bottom level and channel bed level for an efficient flow of the water and complete drainage as and when required. It should be made of a material which can withstand the saline action and where fouling action is minimum.
- Self weight of the gate as far as possible should be lighter but not sacrificing the structural requirements. It should require minimum additional foundation depending upon the soil characteristics and should be economical and easy to construct.
- It should be firmly situated at the bottom of pond or channel and linked with the dikes to prevent seepage and possible collapse.
- must be easy to operate and maintain, and should have proper provision for filtering predatory fishes, their eggs and other organism.

Pond Inlet Structures

Grow-out Pond

- The inlet sluices are meant for supplying the required quantity of water from the supply channel to the pond.
- The sluices may be constructed of brick or concrete walls.
- The sluices must be designed for a peak inflow of 50 percent water exchange. Three numbers of grooves are provided on the inner face of the sluice wall for fixing nylon net and two wooden shutters for controlling the flow of water.
- A nylon net is fixed on the channel side. Whereas, the wooden shutters, one with opening on the bottom fixed immediately behind the nylon net and other on the pond side with opening at the top.
- These two shutters serve the purpose of controlling the quantity of the water flow from 12 to 50 percent of water exchange as and when required.

Rearing Pond

- As the quantity of water to be exchanged is small compared to the grow out pond, inlet sluice made of PVC pipes may be provided.
- The sluices consist of PVC pipes of required diameter partly embedded in concrete and the remaining length of pipe laid down on well compacted soil.
- About 4 sets of grooves are provided on the inner face of wall, two for accommodating shutter and two for nylon nets.
- Anti-seepage collar is to be provided around the PVC pipe for preventing seepage.

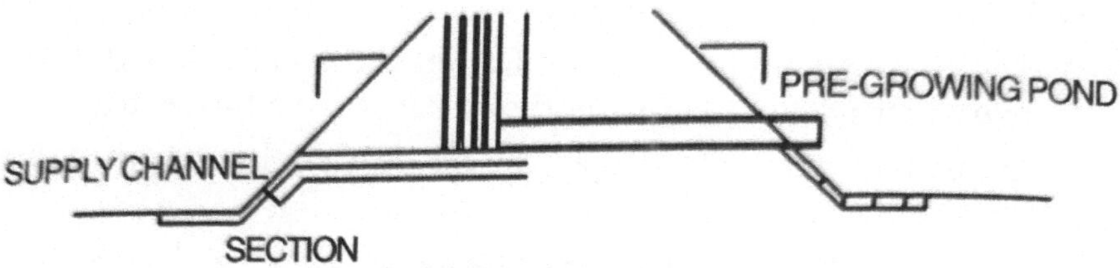

Figure 2: Cross Section

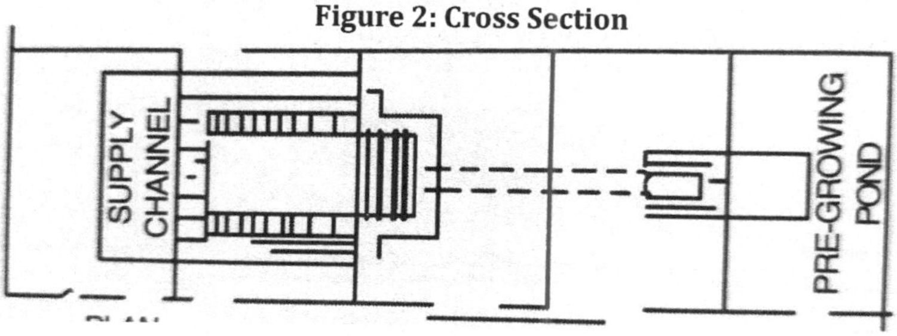

Figure 3: Aerial View Plan

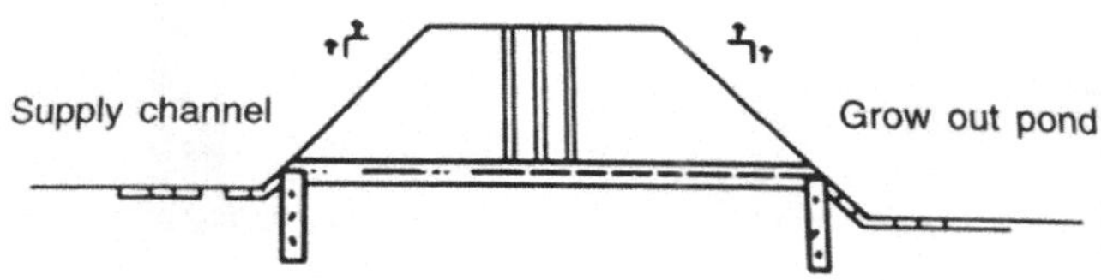

Figure 4: Cross Section

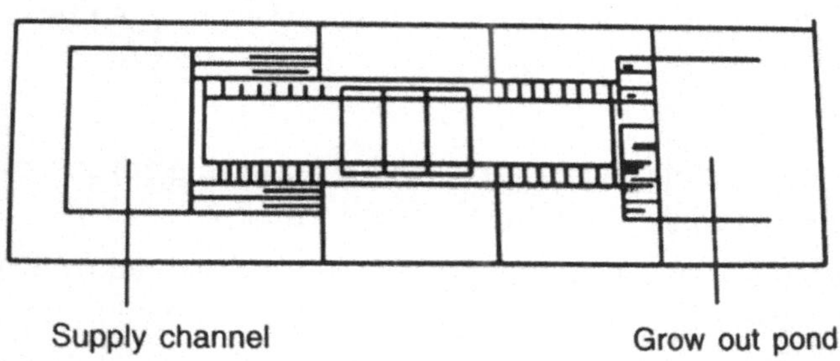

Figure 5: Aerial View Plan

Pond Outlet Structures

Rearing Pond

- The rearing pond outlet is a two-pipe device for draining out water during daily exchange and for complete emptying as well.
- One pipe consists of 25cm diameter of PVC, perforated for about 2m length of the pipe at the bottom, in which a 2nd pipe of 16cm diameter is concentrically placed. This device is used for draining daily exchange of water.

- Complete draining is done when inner pipe is removed.
- The outer outlet pipe of 25cm diameter when placed with the holes towards the bottom, facilitates the exchange of bottom water.
- In case the screen gets blocked, the excess water is discharged through the over flow pipe.
- A walking platform supported on wooden piles may be provided for access to the control device for the sluice.

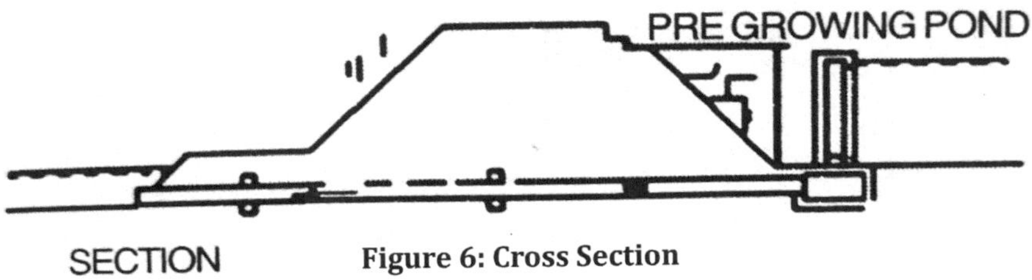

Figure 6: Cross Section

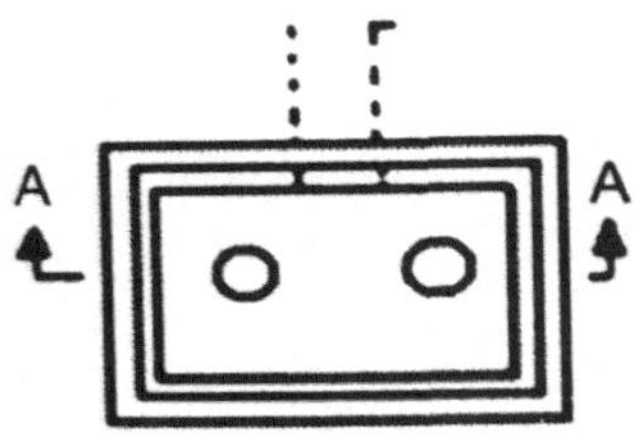

Figure 7: Top View of the Device

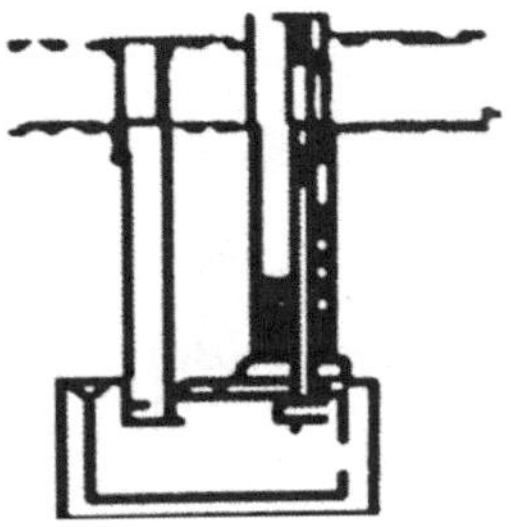

Figure 8: Section A-A (Two Pipe Device)

Grow-out Pond

- The outlet sluice gates are provided for letting out the exchanged water as well as while emptying the pond water.
- The sluice consists of RCC pipe of required diameter. A brick or concrete wall over a RCC raft of suitable thickness is to be provided for housing the control arrangements. Grooves are to be provided on the inner face of the wall for fixing the nylon nets and wooden shutters.
- Cut-off walls are to be provided at the upstream and downstream side end of the shutters for arresting seepage.

Surplus Weir

A surplus weir is to be provided at the end of supply channel with the crest level at full supply level of water in the channel. This will enable the escape of water built up above the full supply level into the drainage channel.

Main Sluice Gate

☆ The main sluice gate connects the main feeder channel of the pond system to the source of water. In a tide fed farm, it regulates the flow of tidal water in and out of the farm. It is the biggest gate in the pond system and is usually made of reinforced cement concrete (RCC). Depending upon the local site conditions and on the size of farm, number of main gates shall be designed. In big aquafarms, There may be more than one main sluice gate.

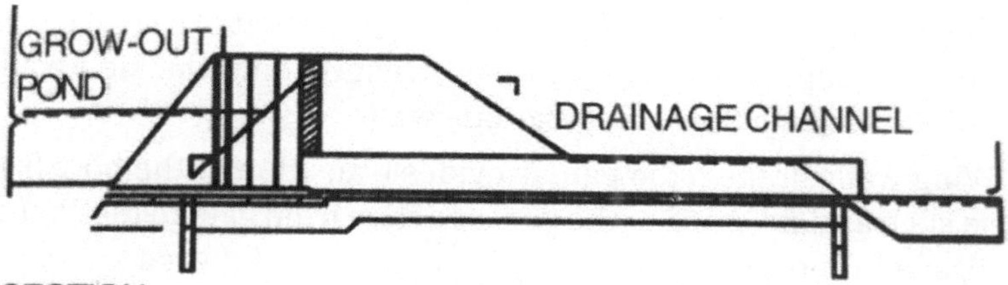

Figure 9: Cross Section

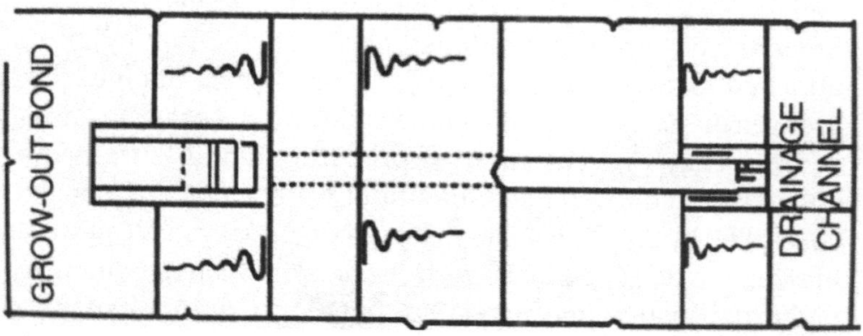

Figure 10: Aerial View Plan

☆ Main gates may have one or more openings with size of each opening usually 1 to 1.2m in width for easy handling. A gate with wider opening will require a lifting mechanism for operation of the gate. The height of the main gate depends on the combined height of highest tide and the river flood. Generally, the height of main gate is taken as that of the main or peripheral dike. It is located at a place where fresh seawater can be obtained most of the time.

☆ The various components of the main gate are floor, apron, cut off walls, sidewalls, wing walls, walking platform, flash boards and screens.

☆ The gate rests on a floor, which serves as its foundation. The elevation of the floor should be lower than the lowest pond bottom elevation inside the pond system of the farm. Preferably, it should be lower than the lowest tide to ensure complete drainage of the system. The gate foundation must be rigid and stable. Often, bamboo poles or coconut timbers are used as piles for providing foundation for the structures. Sometimes layers of boulders and gravel are also used along with the piles.

- The extended and clear part of the floor up stream and down stream of the gate is known as apron. It provides protection against scouring action of flow.
- Cut off walls is necessary at both ends of the gate floor if it lies over alluvial soil, for its safety against seepage and consequent piping action.
- Side or breast walls are used as a retaining wall for the dike fill. In the side walls, grooves are made for fixing screens and flash boards. Most of the main supply gates are provided with 4 grooves, 2 at each side. Usually, the top of the walls made the same height as that of the dike. Sometimes buttress supports are also be for strengthening the side walls and to reduce seepage flow along the side walls.
- Wing walls in the main gate provides transition of the flow from water so such as tidal creek to the gate and then from gate to the channel. They also help is retaining the earth on both sides of the gate.
- Catwalks or walking platforms are provided between the two side walls. Usually they are made of RCC slabs or wooden planks.
- Screens containing high-density polyethylene or nylon meshes are attached to a wooden rectangular frame, which fits into the two outer grooves of the gate one at channel side and the other towards the water source. They prevent entry of predators along with incoming water and exist of cultured fish. Amount of water flowing through the gate is controlled by the flash boards. They are generally made of wooden planks of size 5 cm thick and 30cm wide inserted into the remaining two grooves of the main gate. A one meter wide opening of the main supply gate needs a channel bed of at least 3.0 m wide.
- Design of main drainage gate is similar to the design of main supply gate. However, in case of main drainage gate, usually 2 rows of grooves are made in the side walls-one for mounting wooden shutter and the other for nylon screen shutter.

Outlet Gates Water Exchange Structures

The main function of a pond outlet gate is to regulate the flow of water from the pond into the drainage channel. Outlet gates are designed to perform the following functions:

- To maintain the designed water level in the pond throughout the culture period
- To adjust the water level in the pond as per the requirement
- Flushing out the accumulation of reduced rotten matter *etc.* from the pond by draining the water from the pond bottom

- ✰ Reducing the concentration of algal bloom and excess phytoplankton growth in the pond by accomplishing surface water drainage
- ✰ Complete drainage of pond water by gravity for easy harvest and for drying of the pond bottom
- ✰ Pond outlet gates are constructed in various farms such as open sluices, monk sluices, pipes made structures, *etc.*

Monk Sluices

- ✰ Monks are the most common outlet sluices used in semi-intensive and intensive aqua-farms. Monks are generally constructed either with rich concrete or with reinforced cement concrete (R.C.C.). Monk is an independent structure built in the pond at the foot of the embankment slope and it is attached to the drainage system (*i.e.* drainage channel or drainage underground pipe line).
- ✰ Monks are generally cubical (column like) structures with three of its wall are built of either concrete or R.C.C., which extend from the pond bottom to above the designed water level. The fourth side of the monk is an open face blocked by screen boards and wooden planks shutters to control the flow of water. These screen boards and wooden planks shutters are inserted into the 3 rows of grooves or slots provided in the inner sides of two opposite walls.
- ✰ The wooden framed plastic or nylon screen shutter fits into the outer facing of the pond. It is usually fabricated using two different sizes of plastic or nylon nettings of 6000 or 4000 micron and 700 micron. Coarser netting prevents frequent clogging of the fine netting from weeds and other vegetal matter, while fine netting prevents the escaping of fish and prawn from the pond. Other two grooves facilitate installation of wooden planks or cast iron plate shutters. The drainage of the pond water either from pond bottom or pond drainage can be accomplished by operating the shutter in the central groove. The height and opening of this shutter is adjusted by sliding it up or down or by increasing or decreasing the numbers of wooden planks. The inner most groove mounts the wooden planks which controls the hydraulic gradient and depth of the water in the pond. Its height can be adjusted depending upon the desired depth of water column in the pond.
- ✰ The sluice boards of the monk are 2.5 cm thick and 15-20 cm high. The edges of the boards are notched into half-lap to fit each other better and to prevent the passage of water between them.
- ✰ The monk is attached to a concrete base plates that is at least 1.0 x 1.0 x 0.2m to prevent it from sinking into the sediment.

Monks are Constructed in Various Shapes like-Square Monk

- V-shape
- U-shape
- T-shape
- Y-shape, *etc.*

Advantages of Monks

- Since it is an independent structure not constructed in the body of embankment, vehicular movement can be allowed on the top of dikes.
- They are most efficient in draining the bottom water than open sluice. Maintenance is easy.

Box Type and Pipe Outlets

Box type outlets also called as "furrow type outlets". Rectangular holes are constructed are in the body of embankment with brick masonry or concrete masonry. Flow of water is regulated with the help of screen and wooden planks fitted in the grooves as shown in the figure. Size of this outlet depends upon total flow and rate of water exchange. Generally, such outsets are provided in extensive fish farms.

Pipe type outlet is also similar to that of box type only the change is instead of rectangular outlet, ACC pipes are provided and screens and wooden planks fitted in the grooves regulate the flow of water.

4. Water Budget Equation

Water budgeting is a process of estimation of total quantity of water required for an aquafarm.

Factors to be considered when calculating water budget for an aquafarm are as follows:

i. Individual pond volume
ii. Rate of water exchange
iii. Loss of water due to evaporation
iv. Loss of water due to seepage
v. Loss of water due to transportation in feeder channels
vi. Rainfall in the location

Total quantity of water is calculated by using the following equation

$$Qr = (V_f + V_{rf} + L_e + L_s + L_c - V_{ra}) \text{ m}^3$$

where,

Q_r = Annual requirement of water (m^3)

V_f = Pond volume to be filled (m^3)

= A x h

= cross section area x depth of the pond

V_{rf} = Pond volume to be refilled (m^3)

= V_Fx N (*i.e.* number of refilling x pond volume)

L_e = Water loss from evaporation (m ^{3})

($L_e = A \times E$)

= (pond area x mean annual evaporation)

L_s= Water loss from seepage (m^3)

(L_s = A x T x S)

= (Area x operational time in days x seepage coefficient (m/day)

L_c = Transmission loss in earthen channels (m^3)

= A_c x 1.2 x E

= surface area of feeder canal x 1.2 x annual evaporation

V_{ra} = Water inflow from rain fall into the ponds (m^{31})

= $A_{effective}$ X Ra

(total area of pond x mean annual rainfall)

Water Use and Water Value Index

Water use and water value index for various aquaculture systems and practices may be calculated by using the formulae (Boyd, 2007) as given below:

Water use index (m^3/t) = Water use (m^3)/Production (t)

Water value index (Rs./m^3) = Production (t) X Crop Value (Rs/t)/Water use (m^3)

5. Drainage System

A drainage system is mainly used to empty the pond. The drainage system must be built before the pond walls because some drainage devices go through the walls. The drain pipe should be placed just 25cm above the pond bottom. The easiest way to drain the pond is to place a plastic pipe through the base of the wall/dyke into the middle of the pond. The end of the pipe which is inside the pond should have a screen over it to prevent the fish from entering the pipe. For a 1 acre pond area with

2 m depth, a drain pipe of 15 cm diameter will take approximately 47 hours or a drain pipe of 25 cm diameter will take approximately 32 hours to drain completely.

There are various drainage systems used:

a. Rivaldi Valve

This valve was named after a farmer in Paraguay who first used this system. This is a very good system for one small pond. It is a flexible plastic pipe which is placed on the ground before the wall is built. After building the dyke, turn the pipe up and tie to a stake so that the end of the pipe is somewhat above the usual level of the water in the pond. The pipe is kept tied until it is time to drain the pond. The pipe end must be tied with a screen to keep the fish from going out of the pond.

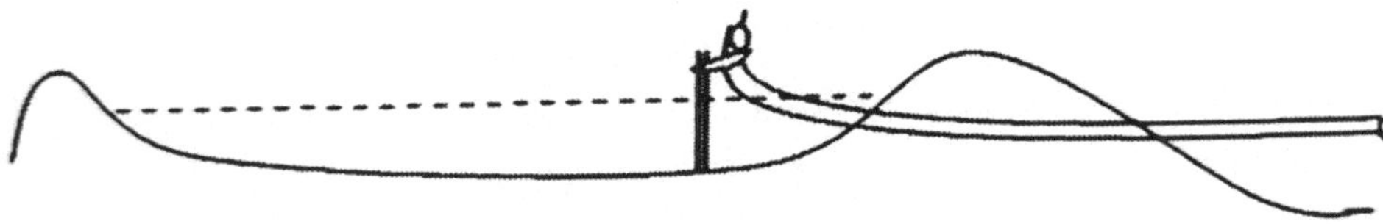

Figure 11: Outlet Pipe and Positioning

b. Elbow Joint

It is a variation of the rivaldi valve. It consists of two metal or plastic pipes connected by an elbow joint. This drainage system is also called as 'turn down' pipe because it is actually turned on its side to drain the pond.

c. Bottom Water Overflow

This drainage system takes water directly from the pond bottom. This regulates the depth of water without any need for moving the pipes. When new water is to be added to the pond, the less oxygenated water at the bottom is drained out automatically.

d. Sluice

Sluice is a type of water control device which can function in a number of ways in a pond. It can act both as an inlet and outlet system. The open sluice gates are the commonly used ones. Sluice is the integral part of the dyke. It is constructed at the midst of the dyke wall before the dyke is built. The sluice can be made of wood, cement or brick. It may have one or two wooden gates. If two gates are used, both the lateral gates (perpendicular to the dyke) should face each other, each provided with two longitudinal grooves for the stop logs or flash boards or vertical lift gates. These boards are removed to empty or fill the pond. A sluice can also have a screen gate to keep the unwanted fish from entering at an inlet and pond fish from leaving at the outlet.

The wooden gates of the sluice must fit into the slots/grooves well but easily. If one board is used in the groove, the entire wooden structure is to be lifted for drainage whereas, when separate boards are used in the grooves, one board may be taken out if a small flow out of the pond is desired.

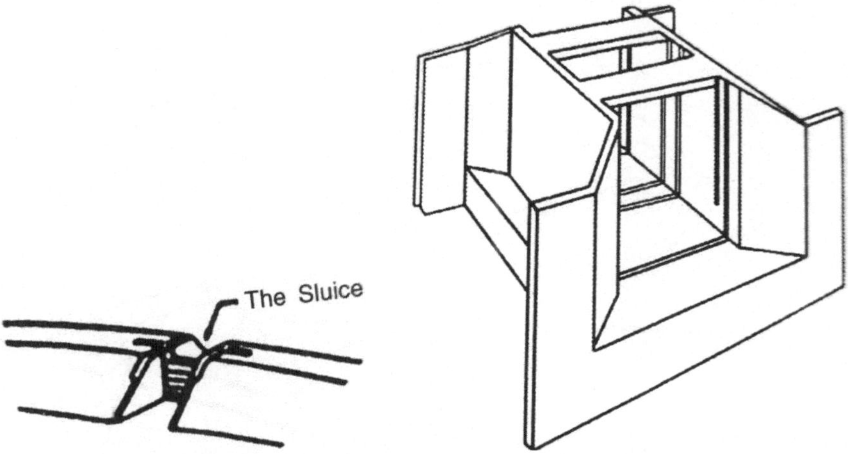

Figure 12: Wooden Gates in Sluice

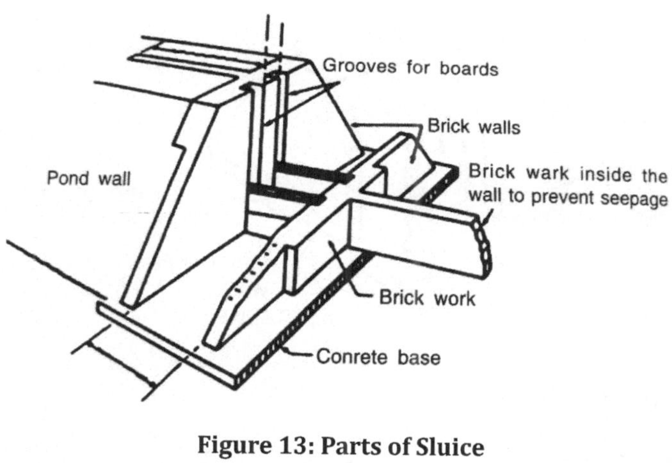

Figure 13: Parts of Sluice

d. Monk

It is the most versatile water control structure used only as an outlet system. The monk is very much like the sluice, but it is not built onto the pond wall. It is a V-shaped vertical tower consisting of the vertical structure, monk and a horizontal drainage pipe. Monk is built in the pond such that the back of the monk touches the wall/dyke.

Monk is a structure which is closed on three sides and open in the front. The open side should be at least 30 cm wide. Entire monk structure should be at least 40 cm above the surface of water. The monk has three grooves. Two grooves are meant for two flash boards and one frame for screen. The boards of the monk must be bevelled or notched to the groove so that they fit together well. At the base of the back wall of the monk is fitted a horizontal pipe which passes through the base of the dyke for draining water.

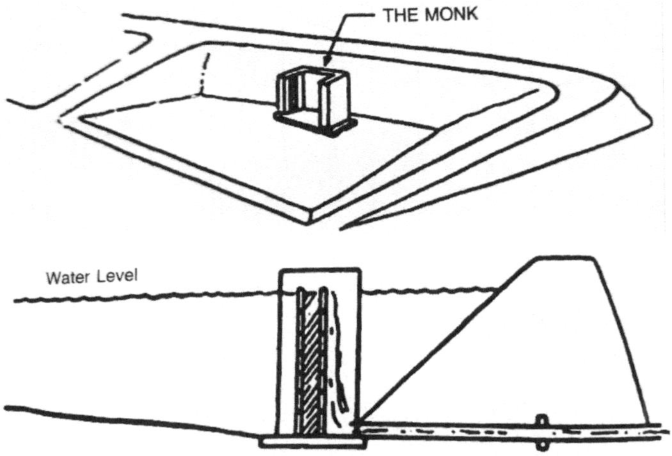

Figure 14: The Monk

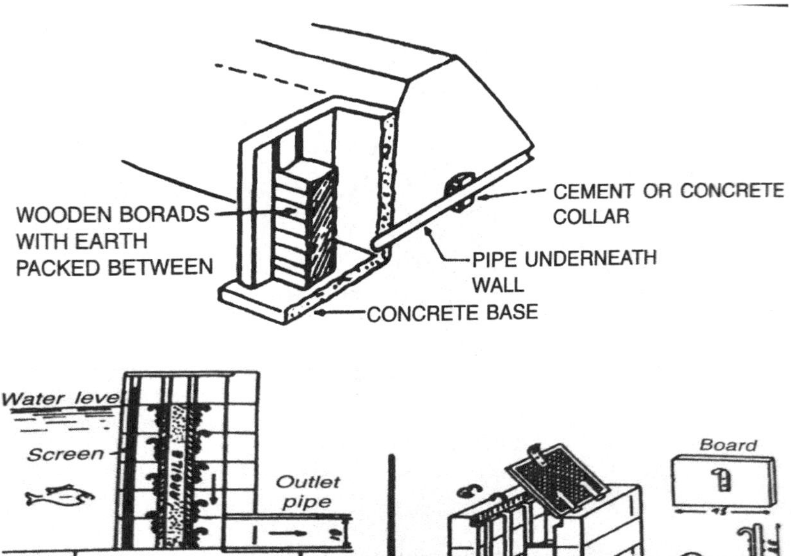

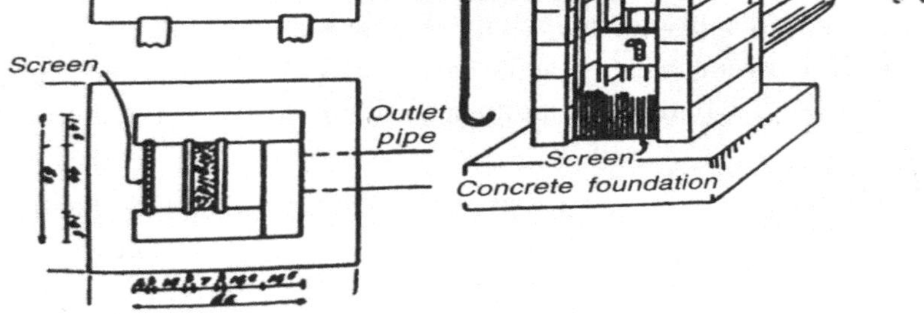

Figure 15: Drain Pipe Design and Positioning

The drain pipe must be placed before the dykes are built. The diameter of the drain pipe should vary between 20cm to 40cm for 1-2 ha to 4-5 ha pond respectively. The drain pipe must rest on solid ground so that pipes do not bend.

f. Herrguth Monk

This is a monk with three grooves. The first groove contains a large screen. The second groove contains a raised wooden gate and the third groove contains wooden gate of flash boards. When a small flow out will be required, one flash board from the third groove may be removed. In this way bottom water will pass through the 2nd gate from the below and then over the 3rd gate from the above. Thus, silt loss from the pond can be prevented.

g. Drainage Ditches

After the ponds are dug out and the dyke *etc.* are built, first of all the pond bottom is compacted and then the bottom is covered with a specially designed network of evacuating channels or drainage ditches. A Fish pond design is recommended for complete and easy dry drainage of the pond.

h. Spill Way

It is a passage built on the body of the dyke of the pond to drain the excess water above the maximum water line when water level will increase due to rain. Such water control structures are of almost importance for such ponds which are susceptible to flooding. It has been recommended that a mechanical spillway is necessary for all stocking ponds with over 2 ha area. Generally, a screen is to be provided at the mouth of the pipe facing inside the pond to prevent escape of fish from the pond. The drain pipe should have sufficient capacity to discharge water for safety of pond.

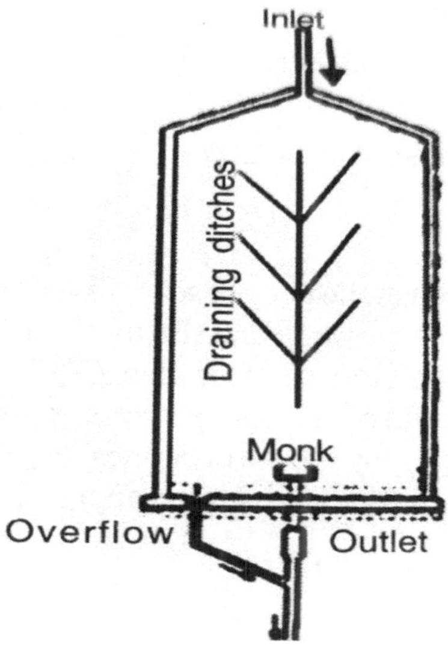

Figure 16: Spill Way

Computation of Discharge through Open Channels

The discharge (Q) through an open channel is given by:

$$Q(m^3/s) = A * V$$

A = Cross sectional area of flow (m^2)

V = Mean velocity (m/s)

The mean velocity is calculated by using different formulae

Chezy's Equation

$$V = C \sqrt{RS}$$

where,

'C' is the Chezy's roughness coefficient

Manning's Equation

$$V = R^{2/3} x S^{1/2}/n$$

where,

'n' is the Manning's roughness coefficient (also known as Manning's constant) For the most economic rectangular sections (for maximum discharges) B = 2d and R = d/2 For the most economical trapezoidal section, R = d/2

- ☆ Flushing out the accumulation of reduced rotten matter *etc.* from the pond by draining the water from the pond bottom
- ☆ Reducing the concentration of algal bloom and excess phytoplankton growth in the pond by accomplishing surface water drainage
- ☆ Complete drainage of pond water by gravity for easy harvest and for drying of the pond bottom
- ☆ Pond outlet gates are constructed in various farms such as open sluices, monk
- ☆ Sluices, pipes made structures, *etc.*

Seepage

The flow of water through the body of soil is called as 'seepage'. The soil plays an important role in holding the water and also to protect from leakage or seepage. If soil has a minimum of 30 percent clay in it, seepage of water through the soil does not take place. For stable soil, the seepage loss is around 0.16 cm/day. Whereas, for permeable soil, it is 0.46 cm/day. In case of permeable soil, pond bottom as well as the walls needs to be sealed. There are several approaches to reduce the seepage.

Methods Used for Seepage Control

i) Clay core

ii) Clay puddle lining

iii) Sealing by polyethylene or plastic or rubber liner
iv) Brick lining
v) Plastic or PVR sheet lining
vi) Clay or biological plaster

i) Clay Core

The clay core is built into the wall and the clay is extended over the bottom of the pond as a lining. This is done when the walls are built.

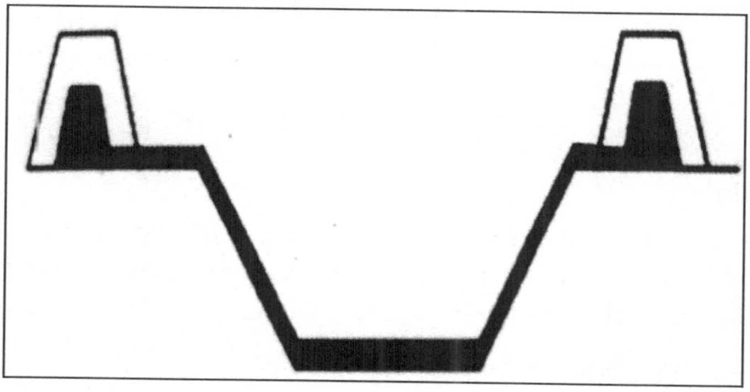

Figure 17: Clay Core

i) Clay Puddle Lining

Clay particles are being used to fill up the gaps in soil and thick layer of approx. 30cm clay puddle layer is put on sides and pond bed which has cementing properties when it remains moist.

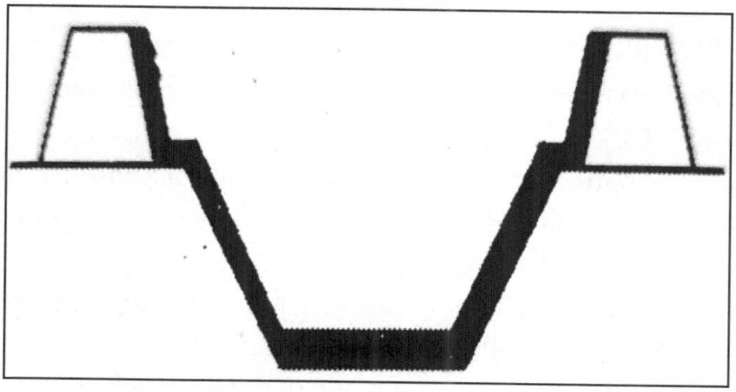

Figure 17: Clay Puddle Lining

iii) Sealing by Polyethylene (or) Plastic (or) Rubber Liner

Generally Polyethylene or plastic or rubber sheet liner are used for the purpose. The waterproof sheet of one piece is placed on the pond bottom and around the sides. Then soil is to be covered over it.

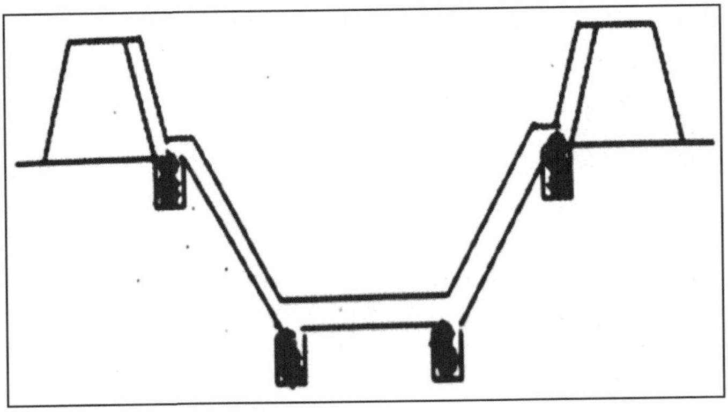

Figure 18: Linear Pond (Fixing)

iv) Brick Lining

It is the most common method. Between two brick layers cement sand in the ratio of 1: 4 and water is put. Over that brick layer, cement plaster is to be made with cement to sand ratio of 1:6. Thickness of the plaster should be restricted to 20mm.

v) Plastic or PVC Sheet Lining

If the soil is highly porous, pond surface is made perfectly even. Plastic sheet is laid right the pond profile and at the corners bricks are laid. If the corner has ground surface, pits are made of depth 60cm and length 45cm and sheet is folded there and sheet is covered with soil. Over the sheet, earth is to be injected with a thin layer of 30cm which will keep the sheet in position. While joining two sheets, a lap of 5 cm should be there and joining must be done using adhesive and putting pressure over that.

i) Gley or Biological Plaster

This method was developed in USSR. The 'Gley' can be made in the pond in the following manner.

- ☆ The pond bottom must be cleaned of debris, rock and all other materials.
- ☆ The pond bottom and sides should be covered with pig dung completely and must be applied evenly.
- ☆ The pig dung layer is to be covered completely with cut grasses, banana leaves or, any vegetable matter.
- ☆ Then a soil layer is to be put on top of the vegetable layer.
- ☆ The layers should be tamped down very well.
- ☆ After 2 to 3 weeks, the pond must be filled with water.

Evaporation

Definition

Evaporation is the process by which water is converted into gaseous stage and returned to the atmosphere as vapour. The water is continuously evaporating from the water surface of natural and artificial sources like river, streams, ponds, reservoir *etc.*, water evaporates from the surface of the soil also. As such a part of the precipitation reaching the earth's surface is lost through evaporation.

Evaporation depends upon several factors such as temperature, humidity, wind velocity, atmospheric pressure, salinity of water and surface area. The rate of evaporation increases with the increase of temperature as well as wind velocity and vice versa. The rate of evaporation decreases with increase of humidity and increases of dissolved salts in water. Evaporation increases when the surface area increases. Rate of evaporation is increases with decreasing of atmospheric pressure. Temperature, humidity, wind velocity. To find out the rate of evaporation commonly used methods are

- ☆ Empherical formula
- ☆ Direct measurement evaporation pan

Factors that Affect the Evaporation

a) **Temperature**: As the temperature increases, the rate of evaporation also increases. Temperature and rate of evaporation are proportional to each other.

b) **Surface area:** As the surface area increases, the rate of evaporation increases. The surface area and rate of evaporation are proportional to each other.

c) **Humidity**: The rate of evaporation decreases with an increase in humidity. Humidity and the rate of evaporation are inversely proportional to each other.

d) **Wind speed:** Increase in wind speed results in increased evaporation. Wind speed and rate of evaporation are proportional to each other.

Erosion of Soil in Dykes and its Control

The water wave developed on the top water surface due to the winds try to notch out the soil from the upstream face and may even some times cause the slip of the upstream slope. Upstream stone pitching be provided to avoid such failures. Heavy rains falling directly over the down stream face and the erosive action of the moving water may lead to the formation of gullies on the down stream face, ultimately leading to the dam failure. This can be avoided by proper maintenance, filling the cuts from time to time especially during rainy season. Grassing the slopes and by providing proper berms at suitable height, so that the water has not to flow for considerable distance. The proper drainage arrangements are made for the removal of the rain water collected on the horizontal berms. Since, the provision of berms ensures the collection and removal of water before it acquires high down ward velocities, the consequent erosion caused by the moving water is considerably reduced.

Chapter 7

Brackish Water Fish Farm

1. Selection of Suitable Site

Sea swells up and recedes every six hours due to gravitation stress between earth and moon. The swelling is at its maximum during new moon and full moon periods occurring at 15 days interval. Such a phenomenon is called "tide", which is responsible for pushing sea water shore ward through river mouths into estuaries, creeks *etc.* causing inland freshwater "brackish", an indication of water characteristic between fresh and saline conditions – hence, "Brackish water Area".

Penaeus monodon (tiger shrimp), *Penaeus indicus* (white shrimp), *Lates calcarifer* (sea bass), *etc.* are abundantly found in these areas. Due to their fast growth rate, accommodating to varied environmental conditions, availability of seed and their overseas high demands, these animals have attracted farmers of brackish water belts as well as Governments of maritime states for their intensive culture.

Adequate knowledge of brackish water aquaculture engineering is one of the prime factors for development of aquaculture farms. While planning for a major aquaculture project, a reconnaissance survey should be carried out by an expert team comprising aquaculturists, aquaculture engineers and environmentalists. The survey may be an aerial survey with the focal attention such as extent of land available for aquaculture, adequate quantity of good quality water source (freshwater, brackish water and/or sea water), the terrain features, soil characteristics, provision of drainage, accessibility to the site, other infrastructural facilities *etc.* Once the team approves the area after reconnaissance survey, for its suitability to aquaculture, a feasibility report may be prepared indicating the project proposal including the species to be cultured, physiological requirement of the species, types of culture systems, targeted production level, fund availability, *etc.*

Hence, selection of a suitable site is one of the most important factor for success of any brackish water aquaculture farm. Generally the survey should be conducted at least during two different seasons *i.e.* summer and monsoon.

Generally extensive, semi intensive and intensive type culture systems prevail for aquaculture farms. In extensive culture system soil and water nutrients are utilized at a maximum level to produce natural food in sufficient quantity which will support for the growth of cultivated species. Mangrove swamps normally satisfy the requirements for an extensive culture system. In case of semi intensive and intensive culture systems, various environmental parameters affecting the growth of animals are controlled. High energy compound artificial feeds are also supplied in these systems.

Criteria for Selection of Suitable Site: In order to select the best suited site for brackish water aquaculture farm, following various factors need to be considered.

i. Topography
ii. Water availability and supply
iii. Soil quality
iv. Environmental factors
v. Power supply
vi. Availability of seed, feed an human resource, *etc.*
vii. Transportation facility
viii. Social and political factors and
ix. Marketing facilities

Out of all these factors, topography, water supply and soil quality are the most important.

2. Construction of Brackish Water Fish Farm

A well designed and laid out pond is characterized by its ease in water management and harvesting and effective predator barriers. The basic requirements of a coastal aquaculture farm are ponds, water distribution and control structures and support facilities. There are basically three types of culture systems *viz.* (i) Extensive, (ii) Semi-intensive and (iii) Intensive. Most of the extensive farms utilize tidal movement for filling in and draining out the water whereas semi-intensive and intensive farms are pump fed farms. In certain cases it may be possible to make use of the tidal movement for water supply in a semi-intensive farm also, provided the following requirements are satisfied.

- ✰ The surface water level in the pond must be at least 0.8m to 1.0 m below mean high water neap so as to get sufficient brackish water.
- ✰ For easy drainage, the pond water level must be at least 0.8m above the mean low water neap level.

In actual practice, it is difficult to fulfill these requirements for tide-fed farms. Therefore, almost all brackish water farms are designed for pump-fed system. The pump-fed system is much more flexible and is easier to operate and efficiently monitored than tide fed farms.

Table 1: Comparision of Pump Fed System vs Tide Fed System

Pump Fed System	*Tide Fed System*
1. Water can be available in sufficient quantity at any time	1. Water intake is based on tidal influence
2. The water is pollution free and free from predators when deep well water is considered	2. Control of predators is difficult as unwanted organisms can enter through the net shutters of sluices
3. Use of pump is cheaper	3. Construction cost is more because of involvement of more earth work
4. Harvesting can be done as and when required	4. Harvesting is possible only during low tide
5. Water in pump fed system will be under control for natural productivity	5. For natural productivity, it cannot be under control

Topography

Topography refers to the changes in the surface elevation of natural ground. While studying the topography of a proposed site, the following points need to be considered. They are:

a) Terrain

- ☆ Terrain is considered most important for site selection. The terrains may be of flat, hilly, sloping or undulating. Although brackish water ponds may be constructed on a wide variety of terrains, slight sloping in one or two directions is ideal.
- ☆ Rocky and sandy terrains are highly unsuitable. Sandy soils cannot be compacted; cause severe slides and are biologically unproductive.
- ☆ Thickly forested terrains are also not suitable as it would cause heavy expenditure.
- ☆ Sites with excessive undulating topography should be avoided since a lot of cut and fill will be needed during the construction and thus would cause heavy expenditure.

b) Types of Brackish Water Areas Available

Several categories of land are found suitable for brackish water farming. They are:

1) *Tidal Mud Flats*

a) These areas get covered with water during thigh tides and are exposed in low tides.

b) Tidal mud flats along with streams are most desirable for location of fish pond to ensure a supply of freshwater when pond water becomes hyper saline and is to be made brackish.

c) These flats do not require any excavation. Constructing a peripheral dyke with a master sluice would convert the area into a farm.

d) These must be cleared of all vegetation.

2) *Swamps and Marshes*

a) Land lying between highly interlacing creeks branching off from the lowermost zones of the estuary is called swamps.

b) These are the areas with low shrubs and other vegetation partially covered with water. Here water table is high.

c) Subsoil is always under standing water which could be stagnant or rising and falling with tide. These areas are rich in organic detritus.

3) *Estuarine Areas*

These areas are the water zone lying between the riverine stretch and open sea with distinct zonation of tides, salinities and water flows. Estuarine areas are broadly classified into 5 zones. They are as follows.

Table 2: Classification of Estuarine Areas

Sl.No.	*Zones*	*Salinity (ppt)*	*Remarks*
1	Head Zone	5.0	River flow dominates
2	Upper Zone	5.0 to 18.0	
3	Gradient Zone	18.0 to 25.0	Most suitable for brackish water farm
4	Lower Zone	25.0 to 30.0	
5	Mouth of the river	=>30.0	Strong tide and sandy bottom

4) *Inlets*

Inlets from the sea and coastal areas are mainly lying on the margin of Bays. There are either covered by scrub or bars. These are partially enclosed by natural shores. These inlets can be made into brackish water farms.

5) *Salt Pans*

- ☆ Disused salt pans are most suitable for shrimp culture.
- ☆ The salinity in these areas is around 40 ppt. During monsoon salinity reduces to 10-15 ppt.

Elevation of Land in Relation to Water Source

- ☆ Elevation of land can be known through land survey *i.e.* contour survey.
- ☆ Lower contour zones should be selected for the construction of ponds.
- ☆ Water source may be utilized either through gravity flow or by pumps.
- ☆ Land which is below the level of water supply suffers from water logging. In this case, embankments will have to be made high to prevent flooding.
- ☆ The higher elevated zones are suitable for infrastructure. Because of higher elevation of the land, the peripheral and secondary dykes need not be high.

c) Economic Considerations

- ☆ Area with high erosion and siltation would pose recurring problems of preventive and protective measures involving heavy expenditure.
- ☆ Desiltation of farm is necessary to keep the required depth of the pond.
- ☆ If site chosen is thickly forested, clearance of lands of roots and stumps would involve heavy expenditure.
- ☆ If site of farm is at a far distant from water supply, maintenance of channel/ canal and its depth requires heavy capital outlay. If water is taken by pumping then reservoir pond needs to be constructed.
- ☆ A financial outlay on road to the nearest transport network is required.

Water Availability and Supply

a) Water Availability

- ☆ Water management is one of the most important aspect of coastal aquaculture.
- ☆ An ideal site must be close to the sea and near to a clean freshwater source or in a region rich in underground freshwater/low saline water. Water should be free from agricultural and industrial pollution as well as turbidity due to high organic load of mud and silt.
- ☆ Brackish water can also be drawn from estuaries and back waters or lagoons throughout the year.
- ☆ A suitable site is one where adequate quantity of clean and pollution free freshwater and sea water should be available to get brackish water of desired salinity.
- ☆ Sea water is the main source of water supply to brackish water ponds. The cheapest sources of freshwater are rivers and lakes. Groundwater or deep well water is the most reliable source of high quality freshwater because it is naturally filtered from unwanted pollutants.

- When pond site is far from the sea, heavy rains can cause the salinity to drop to zero or the heat at the peak of summer can increase salinity over the desired level.
- If the water source from river is loaded with agricultural pesticides, fungicides or industrial wastes, then the water becomes unsuitable for aquaculture.
- If there is silt in the water, it can be taken into account since it can cause a rapid reduction of the pond depth.

A preliminary survey of the water source includes:

i) Flow patterns and directions
ii) Rise and fall in tides
iii) Silt carrying capacity
iv) Rate of evaporation

b) Tidal Range and Elevation

Basic water management practice in a farm involves in replenishment of pond water with fresh tidal inflow as frequently as possible. That is, tidal inflow and outflow of the available source should be such that water can be taken in at every high tide and drained out at low tide.

Supply of tide water at 10 to 15 days interval (*i.e.* spring tides) would be sufficient to keep the brackish water ponds fit for aquaculture.

Biologically speaking, daily exchange of water is not suitable because:

- Supplementary feeding and artificial fertilization would be adversely affected.
- Daily harvesting is not possible (as daily growth is not possible).
- Constant rate of exchange may harm some animals and bring in unwanted predators.
- Economically it is impracticable and unprofitable.

- Sites with average elevation lower than mean lower low water should be avoided as ponds cannot be drained properly.
- Places where tidal fluctuation is high, say over 4 m are also not suitable for pond construction because, very large and expensive dykes would be required to prevent flooding during high tide.
- Areas where tidal fluctuation is between 2 to 3 m are ideal for tide-fed farms.
- Areas with low tidal fluctuation of 1 m or less are unsuitable for tide-fed ponds because the ponds could not be drained or filled properly.

☆ The most ideal site would be having an elevation sufficient to retain the 3 ft depth of water after negotiating with the volume of earthwork necessary for construction of dykes.

Soil

Soils are porous mixtures of inorganic particles, decaying organic matter, air and water and a variety of living organisms. The pH and mineral concentration of soil have a direct effect on the acidity and presence of toxic metal ion in water supplies. Soil texture and structure are important physical properties of soil which affect supply of water and air in soil.

Soil texture refers to the relative proportions of various size groups of particles *viz.* sand, silt and clay in a given soil. Soil structure refers to arrangement of soil particles. It influences the rate at which water and air enter and move through the soil and soil nutrients supply.

Soil classification is normally based on particle size and some soil properties such as plasticity and compressibility. Physical and chemical characteristics of the soil should be critically examined as they decide the economics of pond construction and pond management respectively.

Acid Sulphate Soil

Soils in mangrove swamps often contain high percentage of organic matters. Anaerobic decomposition of such soil behaves like acid sulphate soils. If kept submerged, potential acid sulphate soil will never become acid sulphate. Acidification of soil lowers pH to 4 or less. Therefore, it is important to identity a potential acid sulphate soil during site survey so as to plan the pond construction, leaving such a type of soil unexposed to air.

Chapter 8

Site Selection, Layout, Design and Construction of Brackish Water Shrimp Hatchery

Among the various penaeid shrimps available in the Indian coast, tiger shrimp (*Penaeus monodon*) is the most favoured candidate species for culture. The important cultivable characteristics of the tiger prawn are:

- ☆ Fast growth rate
- ☆ Attains large size
- ☆ High resistance to handling stress
- ☆ High survival rate during culture

The tiger prawn has also good consumer preference in the domestic as well as export market thereby fetching maximum unit value. This encouraged for the production through extensive, semi-intensive or intensive culture systems in the coastal areas and gained popularity in India and other countries. This was accompanied by the establishment of shrimp hatcheries all along the coast by government organizations as well as private companies for assured and regular supply of quality shrimp seeds in time. Although seed required for farming can be collected from the coastal shallows, backwaters, estuarine mouths and tidal creeks, Its availability from the wild depends upon several factors *viz.* spawning success of the stock, survival of eggs and larvae, their recruitment to the seed collection centers and therefore, undependable for planned farming. Hence, establishment of shrimp hatcheries for mass production of shrimp larvae under controlled condition require higher priority to support the commercial farming.

Hundinaga,a Japanese scientist, was the first to successfully breed and rear the larvae subsequently under controlled conditions in the laboratory during the year 1935.

Site Selection

The success of a shrimp hatchery lies on the selection of suitable site for its construction. The site should have the environmental parameters that are essential to operate the hatchery for a maximum period in a year.

The principal environmental considerations for site selection are:

- ☆ Quality of sea water
- ☆ Availability of spawners
- ☆ Availability of electric power
- ☆ Availability of freshwater
- ☆ Easy accessibility

These factors have major influence on the success of the technical operations and the economic viability of the hatchery.

The following are important criteria to be considered for choosing a site for the hatchery:

- ☆ The site should not be selected near river mouths so as to avoid abrupt changes in the water qualities during rainy seasons. Flood water from the rivers also brings a lot of silt and detritus which add to the turbidity of the water.
- ☆ The sea water collected from the area should be clean and relatively free from silt.
- ☆ The sea water should be free from pollution either through industrial or agricultural wastes or through toxic substance producing organisms.
- ☆ The site should not be located near source of thermal, sewage or industrial Pollution.
- ☆ The sea bottom near the site should be sandy or rocky, not muddy. The seashore should be flat or gently sloping.
- ☆ The sea water salinity should be within the range of 28 to 32 ppt.

The desired range of sea water quality for a shrimp hatchery is as shown in Table 1.

- ☆ The freshwater (either from tap or wells) should be available at the site for the staffs daily use as well as to dilute the salinities, as and when required.

Table 1: Desired Range of Seawater Quality for a shrimp hatchery

Parameters	*Range*
Salinity (ppt)	28-32
Temperature (°C)	28-32
Suspended solids (mg/lt)	0-100
Total solids (mg/lt)	0-1000
pH	7.0-8.5
Dissolved oxygen (mg/lt)	8.0-10.0
Unionised ammonia (mg/lt)	0-0.1
B.O.D (mg/lt)	<0.1
Heavy metals (ppm)	<0-0.01

- ☆ Excessive fluctuation in temperature, humidity and seasons are not favourable.
- ☆ Elevation of the land should be relatively high enough to avoid flooding problems.
- ☆ The site should be easily accessible so as to facilitate not only the transportation of hatchery equipments and other related supplies but also to market the seed produced to the farming areas.
- ☆ The proposed site should be situated nearby the prawn farms.
- ☆ The site should also be situated in proximity to a fishing harbor so that adult shrimps (spawners) can be obtained easily.
- ☆ Areas subjected to sea erosion or soil erosion should be avoided.
- ☆ The area should not be affected by cyclones and floods.
- ☆ Electricity in the site is essential to maintain and operate the life supporting systems in the hatchery.

Layout and Design of Shrimp Hatchery

There is no any fixed pattern of layout of the shrimp hatchery. However, the layout of the hatchery should be compact without much foot work for the technicians and easy to operate with maximum efficiency. The arrangement of tanks and working space should be based on working performance of the hatchery and technique to be followed.

Hatchery Facilities

The size of the hatchery depends on the production target and number of days available for hatchery operation. Following are the basic facilities to be provided in a shrimp hatchery.

a) Hatchery building
b) Holding tanks
   - ☆ Maturation tank
   - ☆ Spawning tank
   - ☆ Larval rearing tank
   - ☆ Post larval rearing tank
   - ☆ Algal culture tank
c) Sea water storage and filtration tank
d) Sea water and freshwater supply system
e) Aeration system
f) Power supply system
g) Laboratory equipment

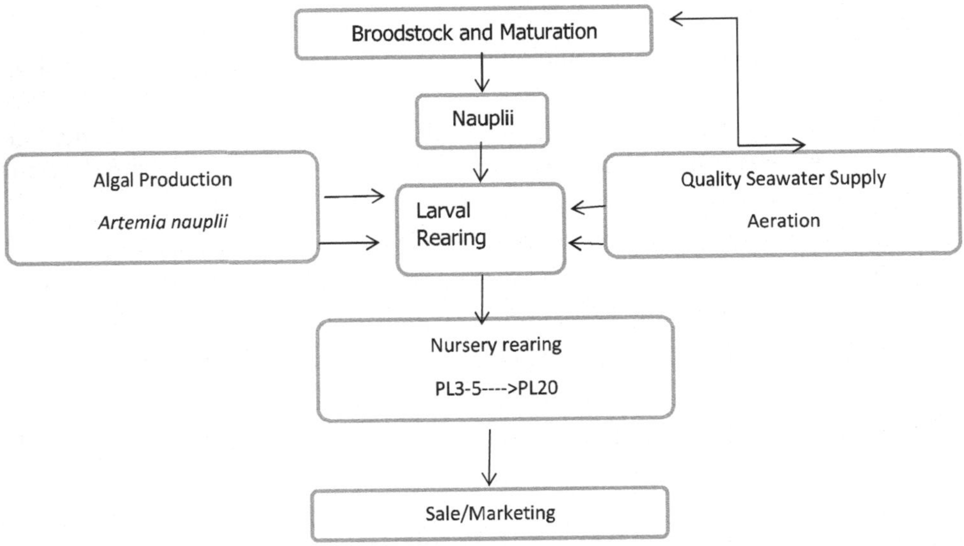

Figure 1: Flow Chart of Hatchery Process

Hatchery Building

The hatchery building is constructed in such a way that all the units are inter connected for easy transportation of live organisms from one unit to the other unit. Some of the units require to maintain temperature. Hence, they should be closed from all the sides. Some of the units require light for algal development. Hence, accordingly, translucent sheets must be provided in between with shedding arrangement below the translucent sheet for restricting the light entry to the required level. Each unit should be provided a chlorine bath before entry to avoid contamination.

The drainage system of the hatchery should be constructed properly, so that there will be no stagnation of water inside any unit to avoid chances of contamination. For this, the entire drainage network should be laid first before starting of hatchery construction to avoid any confusion at a later stage. The drainage system should be planned considering the future expansion.

The wastewater of the hatchery should never be discharged close to the intake point of the hatchery to avoid contamination. The hatchery should also be provided with a laboratory space, space for store, office along with the residential quarters for the essential staff for attending the works during emergency and night hours.

The layout is given to provide facilities maturation, spawning, larval rearing and feed development. Provision for algal stock culture, space for office, garage, machine room, sump and over head tank are also made. Accordingly, the total building area has to be worked out.

A model layout of shrimp hatchery is given in the Figure 2 as follows.

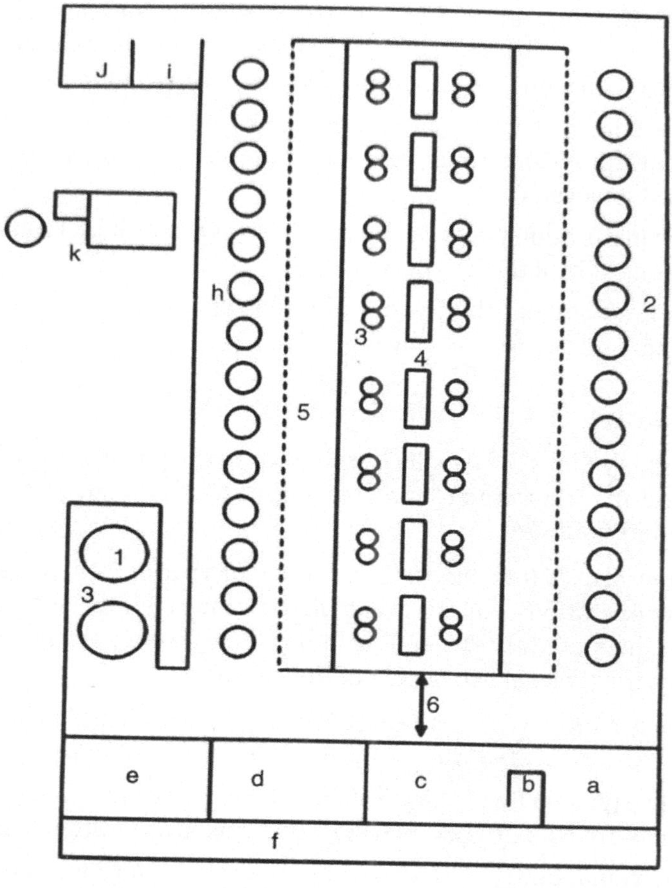

Figure 2: Model Layout of a Shrimp Hatchery

- The main section of the hatchery complex should be a glass-roofed shed.
- In the center of the shed, there should be a 5.0m wide and 2.0m high concrete platform supported on concrete pillars to accommodate the algal culture tanks and spawning tanks.
- On either side of the raised platform, a 80cm wide and 1.2m high catwalk running in the entire length of the hatchery should be provided.
- The larval rearing tanks are arranged along the outer side of the catwalk.
- The glass roofing should be so constructed that the hot air near the roof can escape through suitable air vents.
- The concrete flooring should be provided with proper slope and covered gutters for draining the spilled water and keeping the hatchery clean and dry.
- In most places, the sides of the shed may be protected by chicken mesh or welded mesh to prevent entry of crows and other intruders. But in areas where strong dusty winds prevail during summer months and cold winds blow during winter, the hatchery should have brick walls on the sides with sufficient numbers of windows for ventilation and light.
- One end of the hatchery complex should be asbestos roofed where laboratory, store room, feed preparation room, duty room, toilet, *etc.* must be located.
- The maturation pools can be kept in an asbestos metal roofed enclosure on one side of the hatchery.
- Another shed at the far end of the hatchery should accommodate the generator and the air blower.

Holding Tanks

The hatchery tanks are usually classified as to the purpose for which they are used *i.e.* maturation, spawning, larval rearing, post larval rearing, artemia culture and algal culture tanks.

The material used for construction of the tanks varies from place to place and techniques of operation or on the availability of materials and also on the cost and durability of the materials. Depending upon the scale of operation and longevity of the facility, the tanks may be made of the followings.

- **Plastic lined wooden tanks:** For short term preliminary or temporary operation
- **Fibre glass (FRP) tanks:** Excellent and for permanent operations. These tanks are usually portable and may be used as multipurpose tank.
- **Brick masonary with cement plaster:** Good for permanent operation. Cheap and easy to install under Indian conditions.

- **Reinforced concrete:** Good for permanent operations. Durable and reliable but more expensive than brick masonary.

Tanks made of concrete or brick masonary should be coated with epoxy resin or epoxy paints of food grade to provide smooth interior surface and to eliminate leaching of harmful chemicals into the water and growth of pathogenic organisms due to rough surface. All the corners of the tanks should be rounded off to facilitate proper cleaning and to minimize dead spots in the tanks.

Process of Epoxy Application

Normally 3mm to 6mm (2 coats) epoxy coating is provided on the interior surface. The wall surface should be rough, properly cleaned with freshwater and dried under hot sun before application of epoxy. Then initially the primer is applied over the surface and after drying a coat of epoxy paint is given to the surface. Since, it is costly painting, only experienced and skilled labours should be engaged.

a. Maturation Tank

Maturation tanks are used for the spawners to mature in captivity. Inadequate supply of spawners remains as one of the major constraints in the development of hatchery industry. At present the hatcheries are surviving with the collection of brooders from different coasts throughout India. But as the distance increases, the cost of brooders also increases including transport mortality, *etc.* So, after collecting the brooders, they are transported to hatchery under oxygen packing and then they are acclimatized for 4-7 days in maturation tanks.

- The maturation tanks are circular in shape, measuring 3.6 m diameter, 1.0m height in size and constructed in ferro-cement. These are of 10,000 liter capacity.
- Animals are stocked at a density of 6-7 animals per m^2 at 1:1 ratio male to female.
- The inside surface of the pools should preferably be black in colour to avoid stress to the shrimps.
- For maturation tanks, filtered and disinfected sea water is used. The water should be aerated continuously.
- Sea water is brought in through a vertical PVC pipe (5cm) and distributed through perforated (0.3-0.6mm) concentric PVC pipes.
- The water level in the maturation tank is kept at the level of 60 cm with a constant flow @10lit/min which provides 200 percent water exchange per day.
- Noise in the maturation section should be avoided.

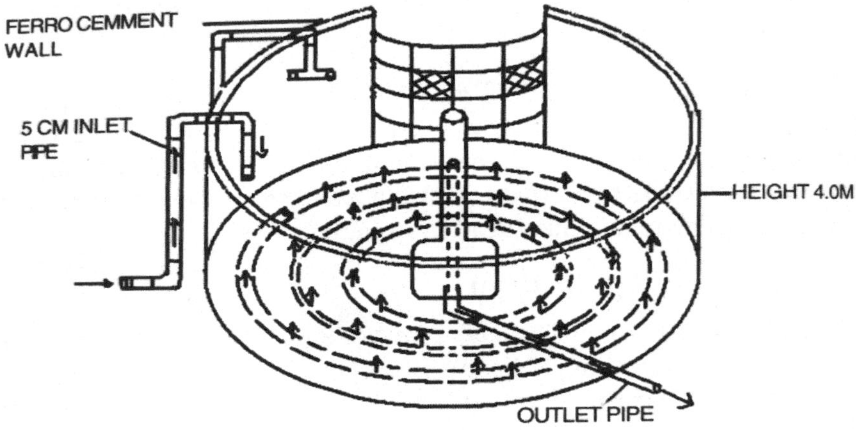

Figure 3: Broodstock Maintenance Tank

b. Spawning Tank

- ☆ The spawners should be spawned in spawning tank.
- ☆ There are cylindro-conical tank of 0.75m diameter with conical bottom having capacity of 250-300 lit which can hold 2-3 spawners at a time for spawning.
- ☆ The tank is provided with a central drain outlet of 2.5cm PVC pipe fitted with a poly propylene ball valve at the bottom. An air stone placed at the bottom of the cone maintains a good circulation of water preventive settlement of the eggs and improving hatching rate.

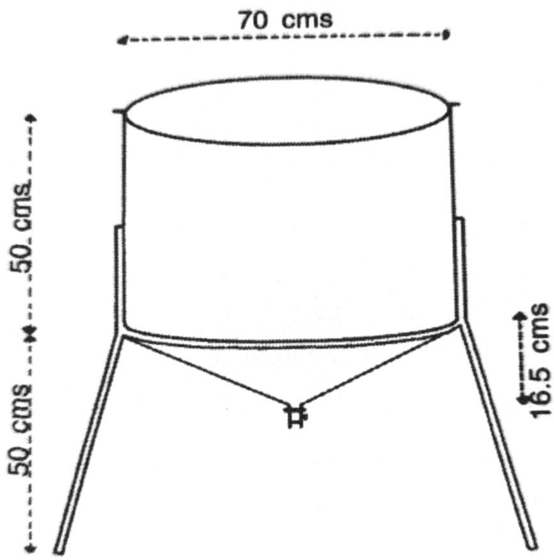

Figure 4: Spawning Tank

- ☆ The tank can either be embedded permanently on the floor or kept on epoxy coated M.S. angle stands of suitable height.
- ☆ The tank can be fabricated in fibre glass.

c. Larval Rearing Tanks

- ☆ The hatched out nauplii are transferred to the larval rearing tanks by siphoning for further rearing.
- ☆ A cylindro-conical tank of 2 tons capacity with conical bottom made of fiberglass is found to be convenient for successful larval rearing
- ☆ It is 1.5m in diameter and 1.0m depth. The conical bottom has a short central drain pipe fitted with a polypropylene ball valve.
- ☆ Two air stones kept at the bottom of the cone keep the larvae and the algal cells uniformly distributed in the tank.
- ☆ The tanks should have a light blue inner surface which is smooth and scratch resistant.
- ☆ The recommended stocking density is 100 nauplii/lit.
- ☆ The nauplii are reared for a period of 13-15 days to attain a size of PL_{3-5}. The expected average survival is about 50-60 percent.

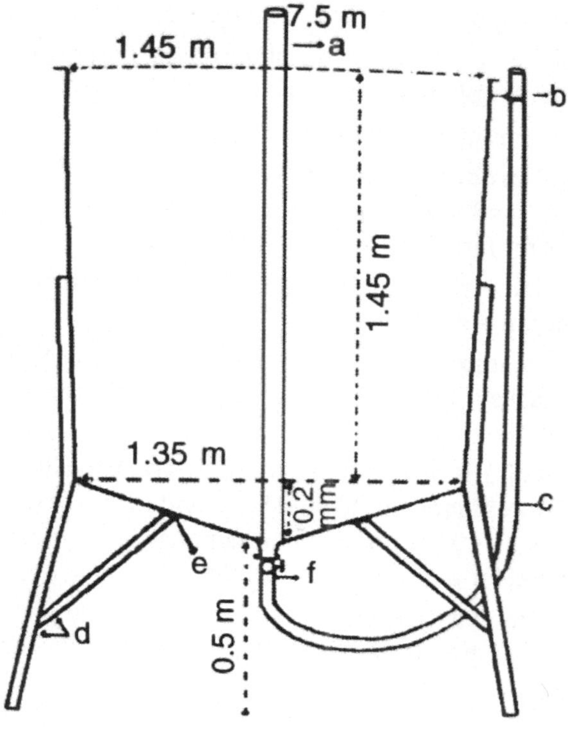

Figure 5: Larval Rearing Tanks

d. Post Larval Rearing Tank or Nursery Tank

- The rearing of PL_{3-5} to PL_{20}, can be done in post larval rearing tank or nursery tanks either in indoor or outdoor.
- The performance of indoor tank is better with reference to survival rate than outdoor tanks.
- The nursery tanks are constructed with RCC or ferrocement or FRP material. The capacity of the tank should be of 10 tonnes.
- The inner surface of tanks should be coated with food grade epoxy paints.
- The stocking of PL_{3-5} is made only at a density of 25 to 50 nos/lit and they are reared till harvest. The water exchange varies from 50-100 percent depending upon the condition of the larvae.
- The expected survival in outdoor tanks varies from 50 to 80 percent.

e. Algal Culture Tanks

- The hatchery requires culture tanks for the feed production of selected diatoms *viz. Skeletonema, Chaetoceros* and *Cyclotella;* flagellates like *Tetraselmis* and *Isochrysis* and rotifers like *Brachionus.* Separate culture of *Chlorella* sp. is essential to maintain rotifer culture.
- A tank of 2000 L capacity with 1.5m diameter and 1.2m height similar to larval rearing tank made of fiberglass is more convenient.
- The tanks should be white in colour on the inner surface to reflect light for enhancing photosynthetic activity.
- Adequare nos of air stones are kept in each tank.
- The algal production can be concentrated at frequent intervals by providing sand filters and stored in frozen condition at pre-determined densities for future use of feeding the larvae.
- For maintaining the stock culture of these organisms, a phycolab is also necessary in the hatchery.
- The main features of the phycolab room are shelves to hold test tubes and conical flasks of 50 ml, 100 ml, 2 L. capacity and carboys of 20 L capacity which are best suited for algal culture.
- All algal culture vessels must be transparent to allow light and easy to wash. Fluorescent lights should be provided for continuous illumination and aeration is to be provided to keep the algae in suspension, to supply CO_2 needed for growth and to stabilize pH.
- Air conditioning is necessary to maintain the temperature around 20°C for optimum growth of the starter cultures as well as to minimize the contamination.

Seawater Supply System

Sea water cannot be used directly for hatchery purpose. The water must be made from suspended particles, bacteria and virus before its use. This needs proper storage and treatment. Generally following treatments are required for sea water.

- ☆ Settlement of suspended impurities
- ☆ Chlorination @5ppm for 6 hours
- ☆ Filtration
- ☆ Dechlorination
- ☆ Filtration to remove suspended impurities upto 1 micron

The sea water supply system consists of:

- ❒ An intake point
- ❒ Pump house
- ❒ Storage tank/sump
- ❒ Overhead tank
- ❒ PVC pipe lines (75 and 50mm dia) to distribute the sea water

- ☆ The water intake may either be from an inshore well or right from the sea. In the later case, extreme care must be taken to avoid entry of fish, jelly fish and other microorganisms as well as silt and mud during heavy runoff. This can be prevented by fitting a removable screen at the free end intake pipe and by employing a sand filter. The pipes are to be made of PVC and should be fully exposed for easy maintenance. Fouling organisms such as barnacles, sponges, oysters, mussels, *etc.* should be removed periodically. Antifouling paints should never be used.
- ☆ In case of groundwater, it should not contain high level of minerals *viz.* sulphur, iron, magnesium, manganese, lead, *etc.* It is essential that the sea water is consistently of the same quality or with suitable chemical parameters.
- ☆ For pumping sea water from the sea to the sump, two (one as stand by) 5 HP electric or diesel pump sets, with open type gunmetal impeller and stainless steel shaft to prevent corrosion are necessary.
- ☆ For pumping water from the sump to the overhead tank two (one as standby) 2 HP electric pump sets with centrifugal gunmetal impeller and stainless steel shaft to prevent corrosion and 50 mm diameter PVC pipe lines are sufficient.
- ☆ The parts that come in contact with sea water should be coated with epoxy paints.
- ☆ If it is possible to install a submerged sand filter at the intake point, the sea water can be pumped directly into the overhead tank, avoiding the

need for a storage sump/tank. Otherwise, the storage tank should be provided with a settling chamber separated from the main storage section by a baffle designed in such a way that the sediment from the tank can be removed easily.

- Water used for hatchery should be clean for which sand filter will be adequate. Capacity for back washing of the sand filter is essential to maintain optimum filtering efficiency. Where more clean water is necessary, filter cartridges along the line can be provided. Usage of ultra violet water purifier will also be effective.
- The daily sea water requirement for the hatchery is to be estimated and accordingly storage tank capacity and that of the overhead tank capacity need to be estimated.
- The overhead tank should have enough height to supply sea water to the hatchery and brood stock tanks by gravity flow.

Freshwater Supply System

Freshwater in the hatchery complex is needed for the following purposes:

a. Cleaning the tanks, hatchery and maturation shed
b. Drinking and
c. Sanitary purposes

The groundwater or tap water can be used for the purpose

Freshwater Supply System Consists of

a. Storage sump (10,000 L capacity)
b. Overhead tank (5000 L capacity)
c. Electric pump sets (1 HP)
d. 40 mm dia pipelines to distribute the water

Air Supply System

- Continuous aeration is essential to maintain the cultures in the hatchery.
- The air supply system consists of air blower/air compressor, PVC air grids, polythene aeration tubes, air stones and regulators. A blower delivers large volume of air at high pressure while a compressor delivers low volume at high pressure. A blower runs continuously while in use and the compressor runs intermittently upon usage as it is equipped with a pressure tank and automatically switches on when the pressure drops below a pre-determined level. Since, higher volume is desirable than higher pressure for the use in hatchery, air blower is preferred.

- Routes type twin lobe air blowers which supply large volumes of oil free air at low pressure are preferred. This is safer and also less complicated to use.
- A blower capable of blowing air through a 2m water column will be adequate for the hatchery. In other words, blower should be capable of supplying air at a pressure of $1 kg/cm^2$.
- Usage of separate small blowers each intended to serve specific area of the hatchery is more preferred than installation of centralized system to serve the entire hatchery complex.
- Air is distributed to the end points through heavy duty PVC pipes fitted with brass or PVC valves and taps wherever necessary, flexible PVC pipes and carborundum air stones.
- Stoppage of aeration is fatal to the larvae and sudden power failure can be disastrous.
- Installation of battery powered warning system would notify the technicians any stoppage of aeration system.

Electricity/Power Supply

- Continuous electricity is essential to maintain the uninterrupted operation of air blower by electric motor and water pump.
- The hatchery also needs electricity. In the area where electricity is a problem, provision of one standby generator or (10-15) KVA is essential.
- High quality rust proof wiring should be used in the hatchery.

Equipment

Various lab equipment are used for regular operation of the hatchery. These are

- pH meter
- Refractometer (salinity)
- DO meter
- Spectrophotometer
- Electronic balance
- Deep freezer
- Refrigerator, *etc.*

Economics

- The expenditure for establishing a hatchery includes capital investment and recurring expenditures. The capital investment includes the cost of land, hatchery building with facilities, for brood stock development, larval rearing, feed developments, aeration and water supply equipments, electrification, office and laboratory equipments, furniture and vehicle.

- ☆ To run the hatchery economically, minimum required staff needs to be included with supporting technical hands.
- ☆ The recurring contingent includes procurement of brooders, feed, chemicals, fertilizers, *etc.*
- ☆ To maintain the hatchery economically viable, it is necessary to run the hatchery for a maximum period in a year.
- ☆ The minimum period for single run of hatchery is as follows.

Table 1: Minimum Period for Single Run Hatchery

System	*No.of Days*	*Total*
Maturation	5 (summer)/10 (winter)	5
Nauplii to PL_{3-5}	13	18
PL_{3-5} to PL_{20}	17	35

Although more number of runs are possible in a year, for working out the cost benefits, a minimum of 8 runs in a year are taken into consideration.

Design of Shrimp Ponds

Three types of ponds are generally constructed in shrimp farms:

1. Pre-grow out ponds (Nursery ponds)
2. Grow out ponds (Stocking ponds)
3. Bio-ponds

Actual size and area distribution for these ponds will be mainly depend upon the culture technology and the function of the farm. Shrimp ponds; especially, semi-intensive ones, require more aeration; and therefore, squarish ponds with low depth (1.0 m to 1.2 m) are ideal. Details of each pond are explained below.

1. Pre-grow Out Ponds (Nursery Ponds)

- ☆ Used for rearing fry up to PL 20 stage, after that they are transferred to the stocking ponds.
- ☆ They are smaller in size (0.1 to 0.5 ha) and will occupy 10-15 percent of the total Productive area.
- ☆ Shape will be either square or rectangular.
- ☆ Nursery pond may be avoided if PL_{20} seeds are obtained from hatcheries.

2. Grow Out Ponds (Stocking Ponds)

- ☆ This is a large compartment in the pond system and keeps the major portion of the total farm areas.

- ☆ Generally stocking ponds occupy about 65-70 percent percent of total productive area.
- ☆ Used for growing the shrimps from PL. 20 to marketable size (18-30 g).
- ☆ Size varies from 1-10 hectares and in few cases up to 20 hectares also.
- ☆ For management convenience, ponds beyond 5-hectare area are not preferred ponds of size 1-2 hectare are ideal; especially, for semi-intensive culture system
- ☆ Although the shape of the pond depends upon the topography of the area,square ponds are preferred from construction and aeration point of view. However in case of rectangular ponds,the ratio of length and breadth will be kept as 1.5:1 to 2:1(L: B).
- ☆ In rectangular ponds, inlets and outlets should be arranged in zigzag fashion. In ponds of size greater than 2 hectare, two inlets and two outlets are provided.
- ☆ The pond bottom should be, as far as possible, smooth and free from projected rocks, tree stumps, *etc.*
- ☆ A gradual bottom slope of 1 in 1000 to 1 in 3000 is essential from the inlet gate towards drainage gate.

3. Bio-Ponds

- ☆ In order to satisfy the pollution control authorities and to prevent environmental degradation, drainage water from the pond should be acceptable to flow to the creek or river system.
- ☆ Bio-ponds are necessary as a part culture pond system.
- ☆ Bio-ponds are large settling tanks where rich nutritional effluents are treated naturally.
- ☆ Also used for secondary culture of fish.
- ☆ Occupy about 7-10 percent of the total area of production.

Recirculating System

- ☆ In re-circulating systems, fish are confined at high densities. Normal culture density is 61-122 kg/m^3, but densities in small-scale experimental systems have exceeded 545kg/m^3.

Economic analyses of recirculating systems indicate two key operating parameters.

i. System operation with minimal energy input

ii. Proper water purification

Aquaculture engineers are challenged to develop designs that optimize these criteria. Contaminants that have negative effects on water quality are ammonia, urea, CO_2, feces and other metabolic waste. Organic wastes are further degraded to produce additional ammonia, nitrite and nitrate. In the close confines of recirculating systems, the effects of these substances are acidic water, depleted dissolved oxygen and increased turbidity, making water more inhospitable to the fish. In addition, many minerals and essential trace elements are lost.

The amount of wastes produced depends on the fish species cultured and life stage, system biomass, and the type and amount of feed given to the fish. The rate and degree of water quality degradation can be managed with proper water treatment. The principal treatment processes include screening, settling (sedimentation), granular media filtration, biological filtration, aeration and disinfection.

Primary clarification includes screening, sedimentation, granular media filtration or a combination of any of the three. It is important to remove solids prior to biological filtration. Biofiltration lowers ammonia and nitrite levels. It is the heart of any recirculating system.

Secondary clarification usually follows biological filtration to remove the biological floc that frequently cloughs from the filter media. It is important that this material is not allowed to remain suspended. Secondary clarification includes sedimentation and may also include screening sometimes. Finally, aeration is added for basic life support system. A water pump drives the system.

Water in recirculating systems completes the entire circuit many times daily. Depending on culture intensity, filtration may be required from two to four times per hour. At a minimum, the water should receive complete treatment one to two times per hour. A daily partial water exchange is necessary to control nitrate, remove pollutants and replenish minerals and trace elements. As much as 90 percent of the water can be recycled daily. The other 10 percent is "new" water. In fact, the total volume is replaced once every 10 days. Lightly loaded systems may fair well with a 3-5 percent daily water exchange. Systems employing 100 percent recycle are rare.

Screening

The simplest and the oldest wastewater treatment process, screening is often used as pre-treatment prior to the primary treatment. Screens placed across the waste stream can trap solids. Coarse ones are used to treat the raw effluent from the culture unit. It can also trap the biofloc that cloughs from biofilters.

Activated Carbon

Activated carbon is made by charring coal, pecan, coconut or walnut shells, wood or animal bones. These materials are heated to about 900°C in the absence of air exposing it to an oxidizing gas at high temperature, then activates the charred material. The gas creates a highly porous structure in the char and a large internal area. Activated carbon can remove volatile organic, color, odor and suspended solids.

Ion Exchange

Certain ions are displaced from an insoluble exchange material (resin) by ions of a different chemical dissolved in the wastewater. Ion exchange resins are manufactured in the form of tiny porous beads about 1 mm in diameter.

Foam Fractionation

This process removes dissolved organic carbon (DOC) and particulate organic carbon (POC) by adsorbing them onto the surfaces of air bubbles rising in a closed contact column. They are also known as protein skimmers. The bubbles create the top of the liquid column and are discarded.

Bio-filtration

This technique utilizes living organisms to remove a substance from a liquid solution. Algae or higher green plants are used. Biofilters can remove ammonia and nitrite. It is generally accepted that, after oxygen, ammonia often becomes the limiting factor in recirculating systems.

Types of biofilters includes submerged, trickling, rotating disks or drums, fluidized beds and low-density media filters.

Submerged filters consist of a vessel filled with a media upon which nitrifying bacteria grow. Culture water is passed through the filter by a spray bar, and gravity pushes water downward through the unit. Completely submerged at all times, the media is supported by a perforated bottom through which the treated water drains and flows out of the unit. Rock is the most common type of media used in submerged filters. Limestone rock is popular since it provides buffering against rapid pH changes

Trickling filters basically function the same way as submerged filters except that the media is not submerged. Culture water trickles downward through the media and keeps the bacteria moist but not completely submerged.

Rotating media filters incorporate aspects of both fixed and suspended media operation. Fixed media are rotated in vessels through which the culture water (suspended media) flows.

In fluidized bed reactors, the media is contained in a vertical vessel with a cylindrical or square cross section. The media is kept in suspension or fluidized by the upward flow of water. The degree of fluidization depends on the upward flux of water through the vessel. Coarse sand is the most common media used.

Types of Shrimp Culture Systems

There are mainly 4 types of culture systems.

1. Extensive
2. Modified extensive
3. Semi-intensive
4. Intensive

1. Extensive System

It is an improved method of traditional shrimp farming with scientific application to improve the quality and quantity of shrimp production. Ponds of size 1-5 hectare are constructed to maintain at least 80 cm of water depth. The range of stocking density will vary from few thousand to 50,000 seeds per hectare without much supplementary feeding and the quality of water is maintained through natural rise and fall of tidal water. Shrimps are cultured for 3-4 months. The production under this system normally ranges from 2 to 3 t/ha/yr in 2 crops. Tide-fed farms are suitable for this type of culture.

2. Modified Extensive System

Ponds of 1 to 2 hectares in size are scientifically designed and constructed with separate inlet and outlet facilities,

- ☆ Stocking density- 0.5 to 1.0 lakh/ha
- ☆ Water exchange- 10-15 percent daily
- ☆ Feeding – Artificial/supplementary
- ☆ Production- 3 to 5 t/ha/yr (2 crops)
- ☆ Aeration- Aerators in emergency conditions

3. Semi-Intensive System

Rectangular or square shaped ponds of uniform size (0.5 to 1.0 hectare) and depth (1.2 to 1.5 meters) are laid out on a well elevated ground having a source of very high quality.

- ☆ Stocking density-1 to 4 lakh/ha
- ☆ Water exchange- 25-30 percent daily
- ☆ Feeding-Artificial/supplementary (high energy prepared)
- ☆ Production-8-10 t/ha/yr (2 crops)
- ☆ Aeration -Paddle wheel aerators or air blowers, 4 to 6 aerators per pond
- ☆ Pump-fed farms are best suited for semi-intensive systems as it becomes very easy to maintain the quality of water. Ponds of size 1 to 2 ha area are ideal for semi-intensive systems.

4. Intensive Culture System

Intensive culture system involves construction of concrete ponds of size 0.01 to 0.1 ha.

- ☆ Stocking density- 5 to 10 lakhs seed/ha
- ☆ Rate of water exchange- 300 percent daily
- ☆ Feeding- Artificial/supplementary
- ☆ Aeration- Air blowers/agitators
- ☆ Production- 20-30 t/ha/yr (2 crops)

These days semi-intensive farming is very popular and is making rapid strides. Intensive farming is very risky due to diseases and other associated problems; therefore, this system is practiced only on limited scale.

Construction of an Aquafarm

Construction is very important step in an aquaculture project. Type of construction mainly depends upon the availability of raw materials like bricks, stones, cement, *etc.* Type of construction also depends upon the function of an individual farm component and culture system. For example, cement concrete ponds are constructed for nursery ponds, ordinary earthen embankments are constructed for extensive ponds, and reinforced cement concrete ponds are constructed for intensive and super-intensive culture practices. The best time period for construction of fish farm is in between late winter and early summer season. Construction process requires careful supervision and skilled workmanship as the success of entire aquaculture project depends upon the efficient construction. Following are the sequence of operations required to be carried out in the process of construction.

1. Land Clearing

The site is to be cleared off from big boulders, stones, trees, bushes *etc.* The land is to be levelled if there is a steep slope or broken surface. There are 3 types of land clearing methods based on the density of vegetation and mode of removal of materials from the site where an aqua-farm is to be constructed.

a) Manual Clearing

If there are only small boulders and thin vegetation like small bushes or trees, then by engaging local labours the site can be cleared off. Clearing the site by engaging labours is called as manual clearing.

b) Mechanical Clearing

If the density of vegetation is thick and complex in nature like big boulders, deep rooted trees, lot of undulations and big ditches, then mechanical equipment like scrapers, bull dozers and other earth moving equipment are engaged to clear the site and this type of land clearing is called as mechanical clearing.

c) Chemical Clearing

If the soil at the site contains harmful bacteria or any other micro-organisms which are harmful to aquatic animals, then appropriate chemicals are mixed in the soil to kill these harmful bacteria, and this method of using the chemicals to kill the bacteria is called as chemical method of land clearing. Chemicals are also used to eradicate unnecessary aquatic weeds,

2. Land Marking

After the land is cleared off, then the land is to be marked for the construction of various farm components as per the layout planning. Dry white lime powder is used to demarcate the positions of various farm elements like dikes, channels and ponds which are to be constructed. It is also called as staking.

3. Excavation

4. Construction of Dikes and Sluices

5. Construction of Ponds

6. Construction of Water Channels and Drainage Units

7. Lining of the Embankment Slopes

The protection of slopes of the dikes against soil erosion is called as lining of the dikes. There are number of methods of slope protection:

a. Stone pitching
b. Brick pitching
c. Cement concrete lining
d. Soil-cement lining
e. Stone slab lining
f. Polyethylene paper lining
g. Growing the grass, *etc.*

8. Hatcheries and other Units

9. Office, Lab, Store Room, *etc.*

10. Construction of Residential Quarters, Watchman Shed, *etc.*

Process of Construction

The first stage in the construction process involves general planning. For this purpose, a topographical map of scale 1:2000 to 1:2500 is required. General features of the farm such as division of the entire farm area into various ponds and their types (rearing or auxiliary ponds), the number of ponds, size of the ponds, the orientation of the ponds, the width of the embankments, and the drainage and water supply systems are worked out on this map.

The next stage involves detailed planning of each pond. For this purpose, a topographic work plan of larger scale is required. This will show where and how much to excavate, and where the earth should be moved. For this purpose a survey has to be made. The pond area is divided into 20 x 20m squares by pegging the ground at these intervals. The peg line should be parallel to the direction of the longest embankment of the pond. If the ground surface between any two pegs is not regular and changes drastically by reason of a ditch or mound, a peg is added at the place of change. These pegs form a network of coordinates in which each peg is marked by a letter and a number. The ground level at each of the pegs and the levels of the peg heads are measured, and the elevations are noted on the work plan of the pond.

Next, the amount of earth required for construction of embankments is calculated from the general plan. This is done by determining the amount of earth required for 1m of embankment (this is equal to the volume of 1 m trapezoidal cross-section of the embankment having the planned dimensions) and multiplying by the total length of embankments. The packing coefficient of the soil should be taken into account.

This coefficient depends on the type of soil and the kind of construction equipment used. The heavier the equipment, the greater is the packing coefficient, usually 20-25 percent This means that when 1m of earth is required for the packed embankment, 1.2-1.5 m^3 should be actually provided.

Generally, about one-third to one-quarter of total pond depth is obtained by excavation. The rest is obtained by depositing earth on the area earmarked for embankments and building them up. The amount of earth which can be excavated is calculated and compared to the amount required (again taking into account the packing coefficient). If these amounts do not balance, tentative excavation depth is changed; accordingly and the amount of earth obtained is recalculated. According to Pruginin and Ben-Ari (1959), 2500-4000 m^3 of earth is required per hectare for construction of pond.

After excavation depth is calculated, the details are transferred to the working plan. The depth of the planned bottom from the head of the pegs is marked. The amount of earth dug out from each square between pegs is calculated, and the place on the embankment where the earth should be deposited is designated. This should be as near as possible to the place of excavation. The contours of the embankment's base are pegged in the field and construction is started according to the working plan.

At the onset of construction, vegetation and roots should be removed. These should not be put into the embankment since organic material will rot and leave behind hollows which become weak points in the embankments and are the first to erode. Generally, top 10 cm soil layer is not used for construction as this soil will form weak bondage between the dike base and ground level. However, since this is fertile and productive, it can be placed on the pond bottom to increase pond productivity.

The drainage pipeline is laid before starting construction of the embankment. The location and length of the pipeline are designed on the working plan. The pipeline should be covered by earth, which is then carefully and methodically compacted by hand tools to prevent leakage along the line. Such leakage may lead to bursting of the embankment in the area of the pipeline. An additional precaution is the construction of an anti-seep collar made of concrete or metal in the embankment around the pipeline and perpendicular to it extending to both sides. The pipeline should be covered with enough earth to protect it against damage from the heavy machinery passing over it.

During the construction work, special attention should be given to the shape of the embankment to make sure that its width at all heights corresponds to the plan. If the embankment is constructed too narrow, this cannot be corrected later.

Chapter 9

Design and Construction of Freshwater Carp Hatchery

India is considered as a carp country. Around 80 percent carp which includes Indian major carps and exotic carps is produced in our country. For development of carp culture, assured fish seed resources is very important. This can only be attained through establishment of carp hatchery. Cultivable major carps generally do not breed in confined waters except common carp although they breed in natural waters such as rivers, bundh type tanks during monsoon months. Looking to its history, natural breeding of Indian major carps was recorded in 1924. Though Indian major carps normally do not breed in ponds and tanks, there were some favourable conditions in certain confined water where they bred during monsoon months when the riverine conditions were stimulated which were termed as 'bundhs'. However, during 1957, Hiralal Choudhury and K.H. Alikunhi discovered the induced breeding of IMC's using pituitary gland for carp seed production in confined waters. Successful induced breeding of silver carp and grass carp was reported during 1955 and 1958 respectively. The circular model of hatchery was developed in China during 1960, which provided suitable environment for induced breeding and egg incubation operations. This hatchery is presently known as ' eco-carp hatchery ' in India and became more popular because of several advantages over other types of hatchery.

Site Selection

Selection of suitable site for establishment of a carp hatchery complex would mainly depend upon the topography, soil condition, water supply and economical aspect of the hatchery.

1. Topography

- Topography of the proposed area should generally be flat or gently sloping towards the outlet.
- The earth available by excavation as far as possible should balance with the earth required for raising the dyke.

2. Soil

- The soil must have enough clay to make the ponds water tight.
- Silty clay or clayey loam is the best soil for the purpose.
- The chemical composition of soil should also be favourable for fish production.

3. Water Supply

- Perennial source of good quality water is highly necessary to fill the pond and maintain water level.
- The quality of water is also important with respect to pH of 6.5-7.5 and DO of 5-9 ppm.

4. Miscellaneous

- Good transportation facility
- Close to source of quality brood fish
- Facility for electricity and communication
- Easy labour availability
- Carp Hatchery Complex

A carp hatchery complex should have the following units:

1. Brood stock unit
2. Main hatchery unit
3. Nursery unit for fry production
4. Rearing unit for fingerling production
5. Packing unit

1. Brood Stock Unit

The brood stock unit is an earthen pond which varies from 0.20 ha to 0.50 ha with a depth of around 2.0m. The pond should have facilities for water supply and drainage to maintain the desired water level. The bottom of the pond should be gently sloping towards the outlet and the side slopes of the dykes may be 1.5: 1 (Horizontal to Vertical).

2. Main Hatchery Unit

The main' eco–carp' hatchery unit for production of spawn consists of the followings:

- ☆ Water supply
- ☆ Breeding/spawning unit
- ☆ Egg collection chamber
- ☆ Hatching/egg incubation unit
- ☆ Spawn collection unit

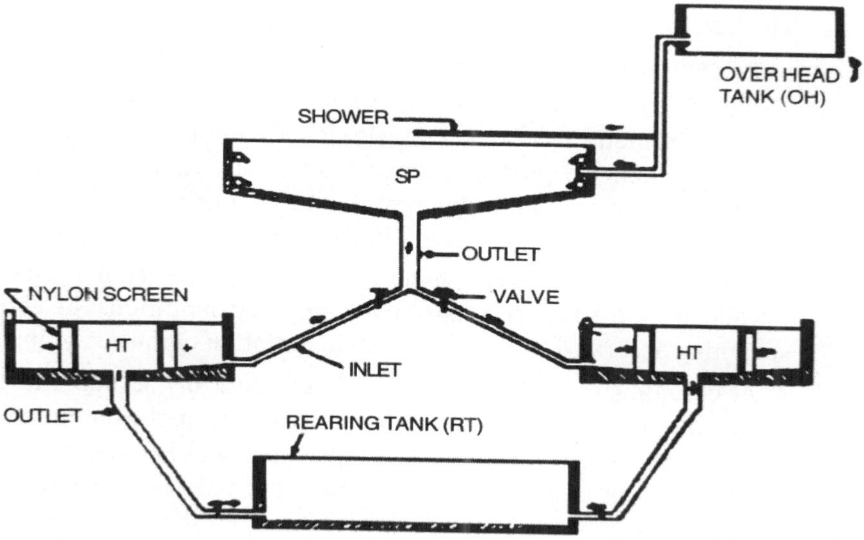

SP: Spawning Pool; HT: Hatching Tank

Figure 1: Cross Section of Chinese Circular Hatchery System

A) Water Supply Unit

- ☆ Most important one for an' eco-hatchery' unit
- ☆ Water source may be of surface water or of groundwater
- ☆ Filtration of surface water is very important which can be accomplished by sand filters. The sand filters can filter particles larger than 0.025 mm size.
- ☆ This will remove most suspended particles including eggs and parasites. Then the water is to be pumped to storage reservoir.
- ☆ When groundwater is the source of water, no filtration is required. It can directly be pumped to storage reservoir.

- In the storage reservoir, the quality of the water (such as pH of 7.0-8.0), temperature of 28-30 ° C and DO of 5-7 ppm) is to be maintained.
- From the storage reservoir, the water is to be pumped to the overhead tank made up of RCC cement and has a capacity of about 5000L.

B) Breeding and Spawning Unit

- The spawning unit is a circular smooth cistern of brick, RCC or FRP
- Dimension: 8-9 m diameter, 0.6-1.0 m deep and 50 m water holding capacity. Water depth is maintained according to the density of brood in the pool.
- The bottom of the pool is sloppy towards the centre.
- At the centre, there is an outlet pipe leading to the egg collection chamber.
- The wall of the spawning pool is provided with diagonally fitted inlet pipes (duck mouths) for circular water flow. Water inflow both in central outlet (egg delivery pipe) and single inlet must be adjusted such that it creates a speed of 3-5 m/s maintaining the requisite water depth. This creates an artificial riverine condition conducive for the fishes to breed.
- About 70 kg male and 70 kg females can be kept in such breeding pool to yield around 10 million eggs in one breeding operation. The effective spawning occurs within 1.0-2.5 hr from initiation of the spawning.
- After breeding, the fertilized eggs will automatically accumulate at the center due to the centrifugal force and pass through the outlet pipe. Provision of showers is essential in ht spawning pool. For this a perforated galvanized iron pipe is fitted above to about half the diameter of the spawning pool. The fine shower from these pipes helps in increasing the dissolved oxygen content of the water

C) Egg Collection Chamber

- The spawning pool is connected to the egg collection chamber through a pipe at the end of which a small net is fixed and the fertilized eggs are collected.
- Then the fertilized eggs will be counted and subsequently transferred to incubation tank.

D) Hatching Unit

- The hatching unit is double circular smooth cistern made of brick, RCC or FRP material.
- The number of these pools in a hatchery system is determined based on the exact requirement. Generally, two hatching units are connected with one spawning pool.

- ☆ The outer chamber diameter varies from 3-4 m and inner chamber diameter remains 1m with a depth of about 1 m and water holding capacity of 9-12m.
- ☆ The ratio of the radius of outer and inner chamber should be 4:1.
- ☆ The inner chamber of the hatching pool is separated by a fine nylon screen which is stretched and fitted on an iron frame. A rubber belt is fastened tightly to seal the compartment from outer chamber.
- ☆ Water circulation is very important for the hatching of fertilized eggs. Some diagonally fitted pipes (duck mouths) act as water inlets at the bottom and the sides of the hatching pool. This helps to move the water around in a circular course and aerating the water.
- ☆ The water speed of the pool is maintained at 4-5 m/s for first 12 hrs, 1-2 m/s for next 6hrs and 3~4 m/s for rest of operation.
- ☆ Based on the running water principle, water enters into the cistern by a series of duck mouth inlets and drains through a screen encircled on the inner chamber.
- ☆ The central outlet pipe being vertically erected maintains water depth in the hatching pool.
- ☆ The hatching time is temperature dependant and varies from 16-20 hrs.
- ☆ The hatching pool can hold from 7,00,000 to 1 million eggs/m^3 of water.
- ☆ During operation of hatching pool, a wooden stick of 2 cm diameter is kept on the water surface or subsurface across the outer chamber in between two walls which is termed as 'surface cleaner' and 'sub- surface cleaner' respectively. This cleaner accumulates the foam and floating debris efficiently and those are removed manually.

E) Spawn Collection Unit

- ☆ It is a rectangular cement cistern placed adjacent or in between the spawning and hatching pool
- ☆ The spawn collection pipe from the hatching pool is generally fixed about 0.3 m above the base of the collection unit to avail a water cushion against mechanical injury of spawn
- ☆ The spawn is collected by gravity flow along with water. Over flow pipe maintains water depth and adequate screen arrangement should be provided to prevent the escape of spawn

Nursery Unit

- ☆ The nursery unit of the hatchery complex is for raising the spawn (4.5 mm) up to fry stage (25-30 mm)
- ☆ These are earthen ponds, small and shallow

- Ponds of 0.02 ha to 0.05 ha size with a depth of 0.7 to 1 m are generally suitable
- Side slopes of the pond dykes should be 1.5:1 (horizontal: vertical)
- Free board should be 0.3 m
- Top width of the dyke should be 1m
- Culture period is 15 days to 21 days
- The ponds should have good quality water supply and drainage system

Rearing Unit

- The rearing unit is used for raising the fry up to fingerling (50 mm and above)
- These are also earthen rectangular ponds slightly larger but need not be proportionately deep
- Size vary from 0.06 to 0.10 ha with depth of 1 m
- The ponds should have controlled inlets and outlets with independent water supply and drainage system

Packing Unit

- Packing unit is an important component in the hatchery complex
- This unit facilitates the seed supply scientifically after due acclimatization and proper oxygen packing in the polythene bags
- Scientific packing practice will reduce the transportation and handling mortality

Chapter 10

Aerators in Aquaculture

Aeration

Aeration is the addition of oxygen to water. Dissolved Oxygen (DO) is the most important critical water parameter in aquaculture system affecting growth, feed efficiency, susceptible to disease and ultimately the fish production. The purpose of aeration is either to increase the concentration of oxygen in the water or to remove gases such as nitrogen (N_2) and carbon dioxide (CO_2) from the water. When the inlet water to a fish farm contains insufficient oxygen or excess of nitrogen or CO_2, there is need for aeration. If the dissolved oxygen concentration in water is low, the culture organisms will not grow well and they will be susceptible to diseases. Too much nitrogen or CO_2 in water is also toxic to fish. Adding excess oxygen to the water will also be toxic for the fish.

There are four major sources of oxygen namely, Diffusion, Water exchange, Photosynthesis and Aeration. Out of all these four sources, aeration is mostly the dependable source of increasing DO in the pond waters.

Dissolved oxygen along with temperature plays an important role in metabolism of fish. Rate of oxygen consumption by fish varies with water temperature, environmental dissolved oxygen, fish size, *etc.* Small fish consume more oxygen per unit weight than large size of same species. Oxygen consumption increases when fishes are forced to exercise and metabolic energy demands. During transportation of fish, the addition of air or oxygen is necessary.

Principle of Aerator

With aerator the aim is to create conditions as near equilibrium as possible between the gas in the air and the gas in the water. Eventually super-saturated gas, especially nitrogen, will be aerated out and oxygen will be supplied if the concentration is below saturation. The aim in constructing an aerator is to achieve optimal conditions for exchange of gas between air and water, so that equilibrium can be reached. By creating a large contact area between the water and the air a good gas-water exchange will occur; see equations for gas transfer and the A/V relationship. The layers of air or water ought to be as thin as possible. To create thin layers it is important that both the current of air and the current of water are turbulent so that effective gas transfer can occur at the water-air interface. The aeration process needs time, and effective aerators will need less time to achieve the same degree of saturation than ineffective ones. Mass transfer from the air will occur into a tank of non-saturated water, almost 100 percent saturation being achieved, but this will take a very long time.

Types of Aerators

Aerators are devices used to increase the concentration of dissolved oxygen in the aquaculture system. There are different classifications of aerators depending upon the design principles employed.

Aerators are classified into 3 types:

a. Surface aerator
b. Sub-surface aerator
c. Gravity aerator

Aerators are classified into 3 types:

a) Mechanical aerator
b) Gravity aerator and
c) Air diffuse aerator

Aerators are also classified into 4 types:

a) Gravity aerator
b) Surface aerator
c) Diffuser aerator
d) Turbine aerator

There are also several aerators that combine two or more types into one unit.

a) Gravity Aerators

- Gravity aerators are simple to construct and are reliable. When water falls under gravity, water gets converted to droplets and the surface area increases *i.e.* the area over which the diffusion can take place is increased and the surrounding air from atmosphere mixes with water. Thereby the dissolved oxygen concentration of the water increases.

☆ Gravity aerators are inexpensive and are most economic when a natural head exists.

1) Perforated Tray Model

In this case the water falls through a screen or perforated plate which helps in aeration. When the height is of Im, aeration of water upto 70-80 percent saturation can be achieved.

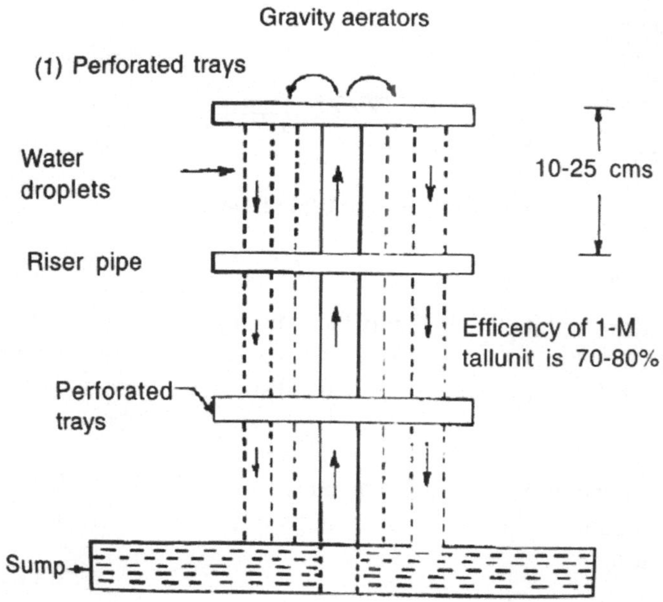

Figure 1: Perforated Tray Model Gravity Aerator (Model I)

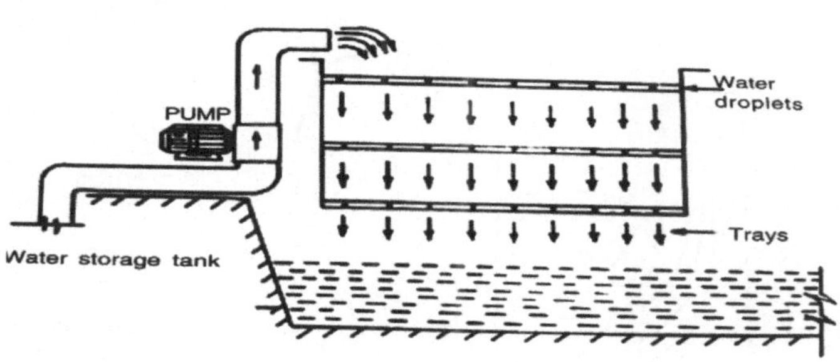

Figure 2: Perforated Tray Model Gravity Aerator (Model II)

2) Gravity Driven Paddle Wheel Aerator

Here the falling water turns the paddle wheel which throws more water into the air. This helps increasing the aeration.

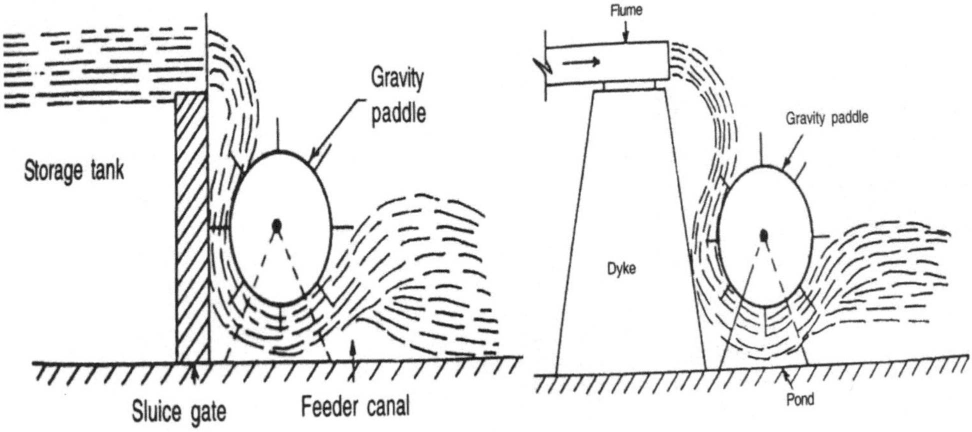

Figure 3: Paddle Wheel Aeration

A) *Surface Aerators*

These aerators agitate the water surface for larger oxygen transfer. By spraying water into the air helps in mixing with air above the water surface which then falls back to the pond. There are 3 commonly used surface aerators.

i) *Nozzles*

In this water is pumped through a nozzle at high velocity. By this a considerable amount of turbulence is created which helps in increasing the dissolved oxygen in the pond water.

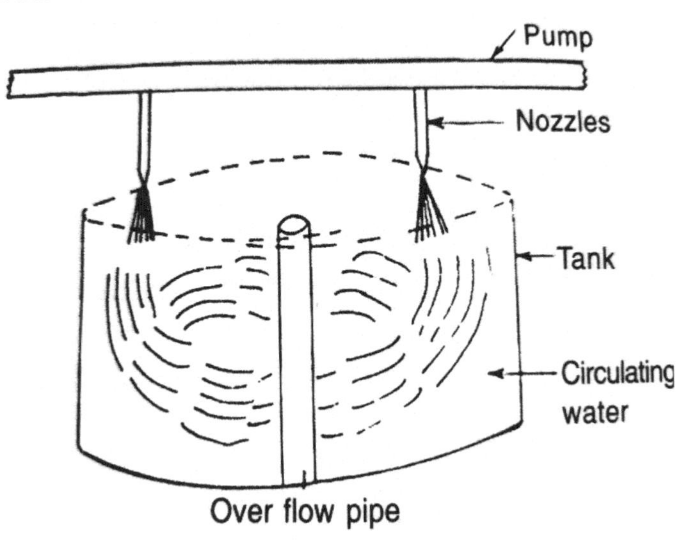

Figure 4: Nozzles Based Water Circulation

ii) *Spray Aerators*

These are the floating propellers placed below the water surface. When the propeller rotates, it creates great turbulence and thus increase dissolved oxygen concentration in the water. The rate of oxygen transfer depends on size and depth or propeller and speed at which it rotates.

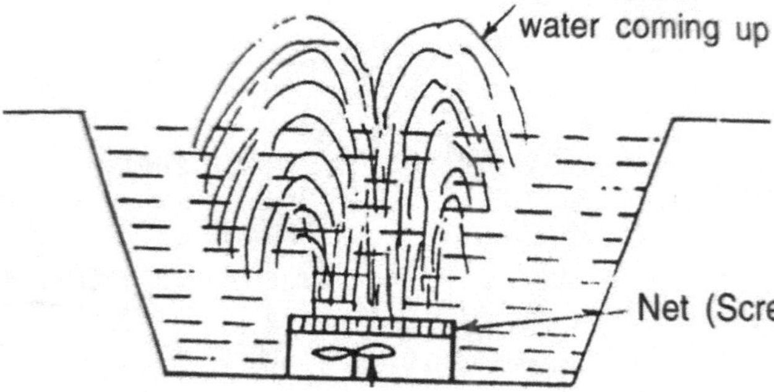

Figure 5: Propeller

iii) *Floating Paddle Wheel Aerator*

These are very popular aerators and one of the most energy efficient devices for increasing DO, concentration. When the paddle wheel of the aerator rotates, the water gets splashed into the air and then oxygenated water falls back into the water. Thereby, the DO concentration of the water increases. The rate of oxygen transfer into the water depends on size and speed of the paddle wheel. The average oxygen transfer efficiency of these types of aerators is 2.13 kg O_2/h (2.13 KW/h).

3) Diffuser Aerators

These inject air or oxygen into the water in the form of bubbles, and O_2 is transferred from the bubbles to the water by diffusion across the liquid film. Since, bubbles rise in a water column, there is relative motion between water and bubbles. This causes water circulation and a renewing of the surface area in contact with the bubble, which increases O_2 transfer. There are various types of diffuser aerators. The simplest is a pipe placed in a tank of water. Air pumped down the pipe forms bubbles at the bottom of the tank. O_2 transfer occurs as buoyant forces cause these bubbles to rise in the liquid.

4) Turbine Aerator

Turbine aerators are also called propeller diffuser aerators. It consists of a propeller submerged in the fluid to be aerated. As the propeller rotates, if circulates the liquid, causing greater aeration to occur at the surface. Oxygen transfer in a turbine aerator depends on the rate of circulation, liquid characteristics and the oxygen deficit in the liquid. Turbine aerators find only limited use in fish culture tanks, since the propeller can easily cause physical damage to the fish.

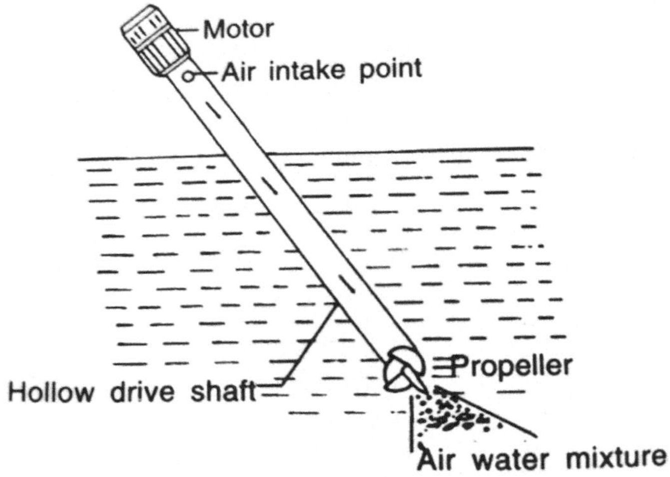

Figure 6: Turbine Aeration

Selection of an Aeration Equipment

Selection of aerator for a pond system depends on:

- ☆ System characteristics
- ☆ Cheap and economically feasible

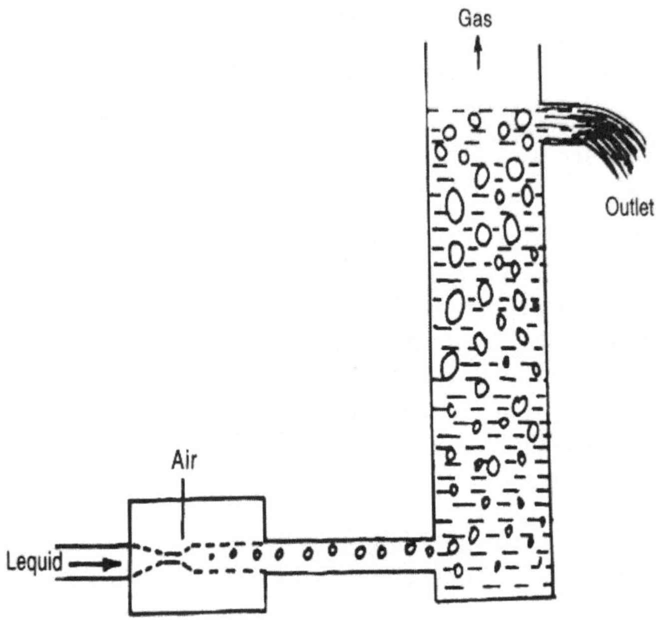

Figure 7: Ventury Aerator – A Diffuser Aerator

- ☆ Easily available
- ☆ Surface aerators are found to have best oxygen transfer efficiencies.
- ☆ Diffuser aerators may be used as airlift pumps as well as aerators.
- ☆ Gravity aerators are most economical choice when sufficient natural head is available

Placement of Aerators in Ponds

- ☆ Proper placement of aerator is important in efficient mixing of oxygen in water throughout pond.
- ☆ Aerators should be placed such that they produce maximum water circulation.

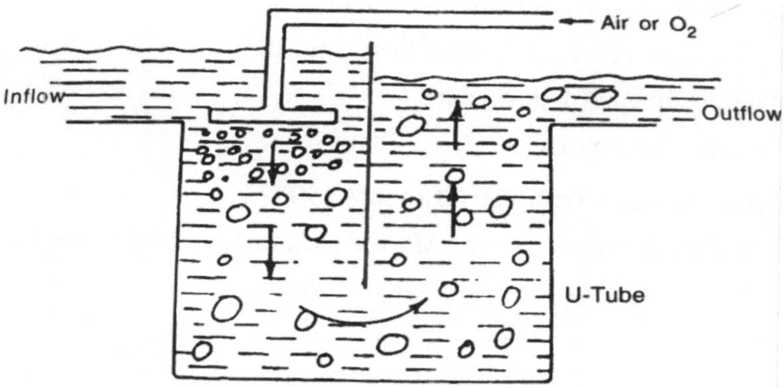

Figure 8: Placement of Aerators

- ☆ For one aerator, location is not critical. It can be placed at any point.
- ☆ For 2 or more aerators, best position is at the middle of one of the long sides of pond directing water parallel to short side of pond.
- ☆ Placement of aerator in the corner of the pond to direct water diagonally across the pond should be avoided.
- ☆ Position of aerators in the pond should be decided based on prevailing wind direction.

Evaluation of Aerator Performance

Aerators are the mechanical devices used to increase the DO content of the water. The performance of an aerator can be evaluated under a rigid set of standard conditions to determine its ability to transfer oxygen. Generally there are two types of testing procedures.

a) Unsteady state test

b) Steady state test

Unsteady Test

Standard tests for aerators are conducted in concrete basins of clean water at standard temperature and pressure (20º C and 760 mm Hg). The water is first de-oxygenated with sodium sulfite and then the change in DO concentration is measured.

i) Standard Oxygen Transfer Rate (SOTR)

SOTR is the amount of oxygen that an aerator can transfer to clean water at 0 mg/lit DO, and at 20° C.

$$SOTR = (K_L a)_{20} (C_s) 20V (10^{-3})$$

Where,

$(K_L a)_{20}$ = Overall gas transfer coefficient at 20° C

$(Cs)_{20}$ = DO, concentration in mg/lit at saturation at t = 20° C

V = Volume of water in test basin

10^{-3} = Converting g to Kg

ii) Standard Oxygen Transfer Efficiency (SOTE)

SOTE is the ratio of oxygen transferred to mass flow rate of oxygen supplied.

$$SOTE = SOTR/m$$

Where,

SOTE = Standard oxygen transfer efficiency

SOTR – Standard oxygen transfer rate

M= Mass transfer of oxygen (Kg/h)

iii) Standard Aeration Efficiency (SAE)

SAE is the ratio of amount of oxygen transferred by the aerator per unit of energy at standard conditions (20° C and 760 mm Hg).

SAE = SOTR/Power Input

SAE = standard aerator efficiency, $KgO_{2/}KW\text{-}h$

Power input = Expressed in KW or hp

Steady State Test

When DO concentration of pond water is greater than zero and water temperature is not always 20° C, this test is applied under field conditions. For testing aerators in the field, two factors *viz.* "α" and "β" must be determined.

$$= \frac{(KLa)_{20} \text{ Pond water}}{(KLa)_{20} \text{ Clean water}}$$

$$B = C'S/CS$$

wherem

C'S = Oxygen saturation concentration for field conditions

CS = Oxygen saturation concentration for clean water

'α' factor is in consumption fluenced by dissolved solids, organic and suspended solids. "β" in wastewater is 0.95 to 1.00 and in fish ponds 0.92 to 1.00 with average of 0.98.

The field oxygen transfer rate:

$$(OTR)_f = (SOTR)_{20} 0^{t-20}\, B(C_s - C_m)./9.092$$

where,

$(OTR)_f$ = Field oxygen transfer rate (Kg/h)

$(SOTR_{20})$ = Standard oxygen transfer rate at 20° C

θ = Theta factor (*i.e.* 1.024 for freshwater)

T = Temperature (°C)

C_s = Amount of saturated DO concentration mg/1

C_m = Amount of measured DO concentration mg/1

9.092 = Value of Cg at 20° C

Commercial Aerators Used in Modern Aquafarms

A) Paddle-Wheel Aerator

It is a surface aerator and It works on the principle of agitating surface water with the help of paddle wheels which are made up of synthetic material. The wheels are on both the sides and a motor is placed in the middle to run the wheel. It uses electricity as the source of energy. The paddle wheel aerators are available with 1 HP, 2 HP and 4 HP motor. The paddle wheels and motor are mounted on the floating frame.

Merits

- It increases the dissolved oxygen level of the water and prevents oxygen depletion at night hours.
- It accelerates the diffusion effect of not only oxygen, but also enables the capture or release of Carbon dioxide as it is important for culture of algae. And therefore for maintenance of appropriate water colour.
- It facilitates the volatilization of undesirable gases such as nitrogen, NH_3, methane and hydrogen sulphide.
- It reduces the daily fluctuation range of pH value.
- It accelerates the decomposition and mineralization of organic matter in water and soil and helps in the release of nutritive value of fertilizers.

- ☆ It diminishes the possible stratification of pH, DO, salinity and temperature in the pond water.
- ☆ It helps in mixing the pond water and maintenance of ideal conditions all over the pond.
- ☆ It increases turbidity when necessary
- ☆ As the pond water moves in a circular fashion, the pond bottom is cleaned and the waste matter gets accumulated in the centre and the corner. By this most of the pond bottom is kept clean.
- ☆ It is very important during application of chemicals or medicine which needs to be distributed throughout the pond and helps to be careful in proper water exchange to reduce any side effect. Use of paddle wheels also has advantage of accelerating the evaporation rate, thereby increasing salinity.
- ☆ Generally 6 to 8 numbers of 1 HP aerators are recommended for 1 hectare pond.
- ☆ Shrimp farm(s) to have proper aeration and regular bottom cleaning. They are kept at four corners and sides and in the middle.

Demerits

- ☆ Paddle wheel aerators waste a lot of energy splashing wastes on the surface.
- ☆ They reach the top 24 to 26 inches of pond, therefore, some sort of DO, pH and salinity stratification may prevail.

Recently the leading companies have brought out paddle wheel aerator with the following features.

- ☆ Motor and gear reducer are specially designed with high quality materials to obtain the high operating efficiency with minimum power loss.
- ☆ Floats and paddle wheels are made of durable plastic (FRP) to assure life with resistance to corrosion and ultra-violet ray of the sun.
- ☆ Paddle wheel consists of detachable propellers which save a slipping space, convenient to assemble and is having low maintenance cost.
- ☆ Rubber coupling is reinforced with 4 layers of synthetic fibre and stainless steel bearing rings. These features assure long life and low maintenance.
- ☆ Base frame and shaft are made of stainless steel which is suitable for all water conditions. Hot dip galvanized frame is also available.
- ☆ Both 1 HP/2 wheel and 2HP/4 wheel type aerator (of multiple of these) are available.

Improved Diffuser Type of Aerator for Wastewater Treatment

It is an improved aerator for wastewater treatment. It can easily be installed. It requires minimal servicing. Operating on the air lift principle, the Ramco MAT-3 aerator creates vigorous subsurface turbulence and exceptionally high oxygen transfer.

Air enters the aerator through a diffuser at its base and is pushed upward. The rapidly rising mixture of air and water is shared by counter-rotating turbine into millions of tiny bubbles. It can be used in many ways from improving oxygen levels and circulation in body of water to cleaning polluted pond bottom to removing any other harmful animals.

B) Aire-O_2 Aerator

A similar version of Ramco MAT-3 aerator which has following merits:

a) Engineered composite material - Rust free, strong and durable
b) Light weight
c) It has power components engineered with only one moving part
d) Improved motor shaft coupling and replaces the universal joint
e) No lubrication required - eliminates contamination
f) It has adjustable shaft angle for operating in shallow and deep ponds
g) Virtually maintenance free saves money in labour and repair
h) Affordable, arrives assembled
i) It is available in 1 HP, 2 HP and 3 HP capacities.

This aerator forces air into water deep down, de-stratifying the pond and enriching it with oxygen. In India, AIRE-O_2 is marketed satisfactorily.

C) Jet-O_2 Aerator

It is a turbine aerator with propeller and diffuser both as described previously.

Chapter 11

Aeration Analytics in Shrimp Aquaculture

In recent years, shrimp outputs per unit of production pond area have steadily increased. Greater feed inputs are made possible by increased rates of mechanical aeration, which are essential for intensification. Paddlewheel, vertical turbine, diffuser and venturi aerators are among the mechanical aerators used largely in aquaculture. In ponds, paddlewheel aerators (of which there are three main varieties) are most commonly utilised. In most aquaculture dominating countries, floating electric aerators with plastic paddlewheels with 1- to 2-hp motors are routinely used in shrimp ponds. Paddlewheels mounted on floats and driven by a power unit located on the pond dike are also used in Asian shrimp farming.

Other types such as long-arm aerators, are powered by 2 to 5-hp electric motors or 7.5- to 16-hp diesel engines. Long-arm aerators are available as prefabricated units from aerator manufacturers, although farmers frequently assemble them from aerator parts. Paddlewheel aerators are sometimes powered by diesel engines. These engines are typically 7.5 to 16 horsepower, normally driven engines that utilise roughly 0.215 litres of diesel fuel per horsepower per hour.

Aeration is measured in horsepower by shrimp growers, but electricity for aerators is measured in kilowatt hours. Farmers were unable to present facts that could be used to segregate aeration energy from total farm energy use.

a. Long Arm Aerators

Based on motor rating information available on the motor specification plates, long-arm aerators usually are powered by 2- and 3-hp, 3-phase, motors supplied with 380-V, 50 Hz electricity from the local grid. The 2-hp motors are rated at 3.6 A with output shaft speed of 1,450 rpm, while 3-hp motors are rated at 5.0 A and 1,420 rpm.

Angle steel frames (typically 4.5 cm x 4.5 cm in cross-section and 0.64 cm thick) are used to install motors. The most popular method of lowering motor output speed is to link a smaller diameter sheave on the motor output shaft to a bigger diameter sheave on the input shaft of a second-hand vehicle transmission using a V-belt. Farmers are able to see a range of transmission brands, however the speed reduction ratios of these devices are usually unknown to them. Power will be delivered from the transmission to the aerator shafts, which are situated on floats 0.75–2.0 m below the transmission output shaft's height.

b. Paddle Wheels

Paddlewheels are usually made of 12–18 paddle pieces mounted on a 5-cm diameter galvanised water pipe. Between the floats or pontoons, three paddle units (1.5 m long, 20 cm wide) will be usually spaced around 0.5–1.0 m apart (about 15 cm in height). Each float would have shaft supports on which the shafts rest. Plastic pillar supports with half-circle indentations upon which the shaft rested are used as shaft supports; however, at one farm, plastic supports with two rollers are used to hold the revolving shaft. The paddle units are arranged in a row, with paddles next to each other. As a result, there are six or eight lines of paddles running the length of the paddlewheel. Each line's paddle tips enter and exit the water at the same time. Nearly all paddlewheel aerators for shrimp ponds have paddles with holes, and there are six or eight paddles per paddle unit. Most paddles are 12–16 cm wide and distal ends of paddles are bent width wise about 5 cm from the leading edge

c. Venturi Aerators

The original Aire-O_2 brand (Aeration Industries, Chaska, MN) or similar devices developed in Thailand and other Asian nations are utilised in tandem with long-arm aerators by the shrimp farmers. Since the late 1980s, Asian-made venturi aerators have been available, however their impeller shafts revolved at around 1,735 rpm, compared to roughly 3,450 rpm for the Aire-O_2 aerators. Slower shaft speeds reduced aspiration, resulting in less efficient oxygen transport to water than the Aire-O_2. Asian-made venturi aerators had 380 V, 3-phase motors with output shaft speeds of 2,850 rpm, and thus they are likely more efficient in transporting oxygen than previous devices with slower shaft speeds.

d. Diffuser Aerator

Long-arm aerators or both long-arm and venturi aerators are also utilised in tandem with diffuser aeration. Roots blowers with air flow capabilities of 3 and 18 m^3/min at 0.3 bar pressure are used to supply air to the diffusers. Blowers blew air into a 15-cm-diameter plastic pipe that ran around the pond's perimeter. Pipes with nipples are tapped at various locations, allowing 2-cm diameter plastic air lines to be extended from main air lines to diffusers on pond bottoms. Diffusers are often created from a farm-made radial arrangement with eight 30-cm long 2-cm diameter soaker hose sections or 3.2-m circular rings of soaker hose.

Aeration Operation Time

Long-arm paddlewheel and venturi aerators are used to create a circular flow pattern in the rectangular-shaped ponds, which had a surface area ranging from 0.275 ha to 0.80 ha. Around the margins of ponds, air diffusers are erected as a regular practice.

In shrimp farms, the most prevalent practise was to run about half of the long-arm aerators during the day and almost all of them at night. During the night, venturi aerators are also used. Few farmers tend to use diffusers running 24 hours a day, whereas other farmers when it was needed.

Energy Loss in Long Arm Paddle Wheel Aerator

The farm-made, long-arm aerator has a 3-hp motor (75 percent efficient) connected by belt drive (95 percent efficient) to an automotive transmission (90 percent efficient).

↓

The transmission is connected to the aerator shaft by two flexible, rubber flange couplings (each 95 percent efficient).

↓

The rotating aerator shaft is supported by six, plastic support bearings (each 97 percent efficient), and the aerator shaft is made of two pipe sections connected by a rigid coupling (99 percent efficient).

↓

The motor outshaft delivers 3 hp, the belt drive reduces power input to the transmission to 2.85 hp (3 hp × 0.95 efficiency).

↓

The transmission further diminishes power to 2.56 hp. Flange couplings reduce power applied to the paddle wheel shaft to 2.31 hp (2.56 hp × 0.9 [2 flanges, 95 percent efficiency each]).

↓

Shaft supports have a combined efficiency of 82 percent (100 percent - [3 percent loss each × 6]) reducing power input to 1.89 hp.

↓

↓

The rigid coupling between shaft sections reduce the power to the rotating paddles to 1.87 hp

↓

The electricity input to the motor is equivalent to 4.02 hp (3 hp × 1.34 kW electricity needed for 1-hp motor output)

↓

The overall efficiency of electricity applied to the motor for operating the paddlewheel is potentially about 46.5 percent

Energy Loss in Paddlewheel Aerators

A 2-hp floating, electric paddlewheel aerator has a motor with an efficiency of about 75 percent connected to a two-directional output, gear box (85 percent efficient)

↓

An aerator shaft is coupled to each gear box output shaft with a rubber flange coupling (each 95 percent efficient)

↓

Ends of aerator shafts with paddlewheel units are supported on each end by plastic shaft supports (each 97 percent efficient) mounted on a metal frame attached to floats

↓

Working through energy losses as performed above for the long-arm aerator reveals an overall efficiency of energy use of 53.4 percent for the floating, electric paddlewheel aerator—somewhat greater than the potential efficiency of the farm-made, long arm aerator

Comparison of Simple Paddle Wheel Aerators and Long Arm Paddle Wheel Aerators

Long-arm aerators fabricated in the farm are less efficient than projected potential efficiency because misalignment of belts and pulleys, which could result in up to 20 percent energy loss. Unbalance and vibration in aerator driver shafts observed at the farms would further increase energy loss. Observed misalignment of aerator shafts on their supports because of instability of floats would cause energy loss. The actual drive train efficiency of farm-made, long-arm aerators would be difficult to measure, and each would likely have a different efficiency

Energy Budgeting of Aerators

Penaeus vannamei is the most common shrimp species produced in aerated ponds, however the same principles apply to aeration of both *Penaeus monodon* and *P. vannamei* ponds, as well as ponds for fish production. Energy use data on pond aeration can be applied across species and especially across Penaeid shrimp species.

Data from Boyd *et al.* (2017) and Boyd, McNevin, and Davis (2018) permitted estimates of average energy use per tonne shrimp from average crop yields, average installed aeration capacity, and average hours of daily aeration during crops in India, Thailand, and Vietnam. The average aeration energy use of 23.4 GJ/t shrimp in Vietnam, 29.4 GJ/t in India, and 29.9 GJ/t in Thailand tends to be greater than that reported in the other studies mentioned above. This likely resulted because the surveys were made during a period when disease problems had been common resulting in lower-than-normal survival and yields. Such is also seen in the study by Mohanty (2001) in which average energy use was 21.8 GJ/t, but if the crop with low survival and yield is omitted, average aeration energy use for the other crops was 15.1 GJ/t shrimp. Problem in shrimp aquaculture, and when individual ponds or entire farms are affected by disease, aerator energy use will increase because low survival leads to lower yield.

Data from Mohanty (2001) reveal that a longer crop results in greater energy use per tonne of shrimp for aeration. It is not clear whether this results from larger shrimp growing slower than smaller shrimp and thereby increasing energy use for aeration disproportionally to final yield. Longer crops may result from shrimp growing slower because of impaired water quality to include water temperature outside optimum range. Obaldo and Tacon (2001) reported that larger shrimp use feed less efficiently than do smaller shrimp, and this could lead to more waste and water quality deterioration during extension of the crop period necessary to obtain larger shrimp.

Aeration Capacity vs Shrimp Yield

Sandifer, Hopkins, and Stokes (1987) suggested a y/hp ratio of 600–750 kg/hp, which was similar to what had been applied in the 1970s in intensive shrimp farming in Taiwan (Hopkins, Stokes, Browdy, and Sandifer, 1991). Wyban, Pruder, and Leber (1989) reported a y/hp ratio of 570 kg/hp, and Fugimura (1989) recommended 500 kg/hp. Hopkins *et al.* (1991) clearly confirmed by research that a ratio of 550–600 kg/hp was usually sufficient to prevent excessively low dissolved oxygen concentration, but this y/hp ratio was occasionally not sufficient to avoid early morning dissolved oxygen concentrations below 3.0 mg/L.

A guideline for installed aeration capacity often given in manuals on shrimp culture is to have 1 hp of aeration for each 400–500 kg increment of the expected harvest biomass. However, this guideline is primarily derived from results obtained with factory-manufactured 1- or 2-hp, floating, electric paddlewheel or venturi aerators (Boyd, Torrans, and Tucker, 2018).

An aerator in practical use will not transfer as much oxygen as suggested by its SAE (Table 1). The main reason is that the rate of oxygen transfer decreases rapidly as dissolved oxygen concentration increases. At a pond dissolved oxygen concentration of 3.0 mg/L, actual oxygen transfer efficiency will be about half of measured aerator SAE (Boyd, 1998). Oxygen transfer rate also decreases slightly in response to greater salinity or greater suspended solids concentration, and it increases with higher water temperature (Stenstrom and Gilbert, 1981). The important points are:

(a) An aerator with a greater SAE will transfer more oxygen than will an aerator with a lesser SAE under identical conditions;

(b) As dissolved oxygen concentration steadily rises during daytime in ponds because of photosynthesis, aerators become progressively less efficient in transferring oxygen;

(c) No net oxygen transfer occurs when water is oxygen-saturated;

(d) Aerators transfer oxygen from water to air when water is supersaturated with dissolved oxygen.

Table 1: Typical Ranges in Standard Aeration Efficiency (SAE) of Common Aquaculture Aerators [Modified from Boyd, Torrans, and Tucker (2018)]

Types of Aerators	*SAE (KgO$_2$/hp-hr)*
Paddlewheel1- to 2 hp floating, electric (Asian type)	0.9-1.2
10- hp floating, electric (U.S. catfish pond type)	1.4-1.8
Venturi	0.9-1.2
Vertical Pump	0.8-1.1
Diffuser	0.4-0.8

Asian, floating, electric paddlewheels, venturi, vertical pump, and diffuser aerators did not differ greatly in their SAEs (Table 1). Shrimp ponds are only 1–2 m deep, and bubbles rise to the surface from diffusers too quickly for efficient oxygen transfer (Boyd and Moore, 1993). In shrimp ponds, diffuser aerators will be less efficient than indicated by SAE which usually is determined in basins with water depths greater than those of shrimp ponds. Tanveer, Roy, Vikneswaran, Renganathan and Balasubramanian (2018) also reported that small paddlewheel aerators used in shrimp ponds had SAE values around 1.0 kg O_2/hp-hr.

Suggestions have been made for improving the oxygen transfer capacity of aerators used in shrimp farming, for example, Boyd (1998), Bhuyar, Thakre, and Ingole (2009), Bahri, Setiawan, Hermawan, and Zairin Jr. (2017), Roy, Moulick, and Mal (2017), Boyd, Torrans, and Tucker (2018), but little progress has resulted. Shrimp pond aeration is still accomplished mainly with small, floating, electric paddlewheel aerators or long-arm aerators, many of which are farm-made. Farmers usually consider factory-made, long-arm aerator as being too expensive.

Role of Aerators in Water Circulation of Shrimp Ponds

Aquaculture aerators and paddlewheel aerators in particular have been designed primarily for dissolved oxygen transfer. Aerators also circulate pond water, and this circulation has several effects. Flow of water away from the aerator avoids formation of a zone in which dissolved oxygen concentration is near saturation and impedes the effectiveness of the aerator to transfer oxygen. Water circulation also mixes dissolved oxygen within the pond volume and moves oxygenated water across the pond bottom which is particularly important in the case of shrimp that spend much time at the pond bottom. While water circulation is beneficial, it also has negative aspects.

Fish and shrimp must maintain their position in a current, and respiration rate tends to increase exponentially with greater water velocity (Ohlberger, Staaks, Van Dijk, and Holker, 2005). Zhang, Zhang, Li, and Huang (2006) found that *L. vannamei* experienced fatigue at water velocities above 5.41 cm/s. Dai, Zhang, and Zhang (2008) reported that shrimp growth increased at water velocities of 0.63 to 2.78 cm/s, but growth decreased at higher velocities. Wijesekara, Nomura, and Matsumura (2005) found that in small tanks, shrimp congregated in areas where water velocity was low.

The influence of paddle shape on energy use and SAE of paddlewheels was studied by Ahmad and Boyd (1988). A paddle bent along its lengthwise centreline in a metal press to form a 135° interior angle in the paddle face was more efficient than flat paddles, paddles with other interior angles, or paddles with bent tips. Friction created by water passing through holes in paddles reduced paddlewheel SAE below that realized for paddles of the same shape but without holes. Four paddles per paddle unit resulted in a greater SAE than either six or eight paddles per paddle unit. The SAE varied also with paddle tip depth as well as paddlewheel speed and diameter. The amount of oxygen transferred increased with deeper paddle tip depth and greater speed (at same paddlewheel diameter), but energy use increased at a greater rate to reduce SAE. There are practical limitations on paddlewheel length, which impose limits on the application of this relationship. Also, water circulation is beneficial in aerated ponds, and water circulation is improved by greater paddle tip depth. Paddlewheel speed for optimum SAE increased as a function of decreasing paddlewheel diameter showing that the angular velocity of the paddle tip was a factor controlling oxygen transfer.

Consider a 2-hp Asian floating, electric paddlewheel aerator (SAE $\approx$ 1.1 kg O_2/hp); a farm-made, long-arm aerator (SAE $\approx$ 0.75 kg O_2/hp); and a U.S.-made catfish pond aerator (SAE $\approx$ 1.35 kg O_2/hp). The Asian made paddlewheel is 1.46 times more effective per horsepower than the long-arm aerator, while the catfish pond aerator is 1.22 times more effective than the Asian paddlewheel aerator and 1.8 times more effective than the longarm aerator for aeration.

The cost of aerators per horsepower is (FOB from manufacturer) Rs.18500–22000 for a 2-hp floating, electric paddlewheel; Rs.74,000–88000 for a 16-hp diesel-powered, factory-made, long-arm aerator; and likely around Rs.7300/hp for a farm-made, long-arm aerator.

Staging Operation of Aerators

Shrimp post larvae for stocking are seldom larger than 0.5–1.0 g and usually much smaller. Shrimp biomass and daily feed input are not great during early weeks of culture. Dissolved oxygen demand also is much less early in the crop, but it increases steadily until harvest because of greater daily feed input and shrimp biomass. Greater nutrient availability from uneaten feed and shrimp excrement enhances phytoplankton abundance as grow-out progresses. Phytoplankton provide dissolved oxygen as a by-product of photosynthesis, but they also consume considerable oxygen and their decaying remains stimulate microbial decomposition which also requires dissolved oxygen.

There is a daily cycle in dissolved oxygen with highest concentrations found in daytime when photosynthesis is occurring, and lowest concentrations observed at night while the entire pond biota (including phytoplankton) continues to respire in the absence of photosynthetic oxygen production. Ponds usually are at or near dissolved oxygen saturation or even supersaturated with dissolved oxygen from 10:00 a.m. to 4:00 p.m. suggesting no need for aeration. However, shrimp live on or near the pond bottom, and some aeration during daytime is desirable by maintaining the circulation of oxygenated water over the bottom. Early in the grow-out period, less aeration is necessary because of lower oxygen demand. Nevertheless, many shrimp farmers run half of the entire installed aeration capacity or more during daytime throughout the entire crop period, and at night all aerators usually are operated. Mohanty (2001) clearly demonstrated that aeration in shrimp ponds could be staged by incrementally increasing the proportion of installed aeration in use and the number of hours per day that this proportion of the aerators is operated as grow-out progressed toward harvest. He measured dissolved oxygen concentrations to confirm that this methodology did not result in undesirably low dissolved oxygen concentration. He presented guidelines for staging aerator operation based on 16 hp/ha of 1 and 2 hp floating, electric paddlewheel aerators for *P. monodon* crops for 110 days or more. This guideline was modified for use with other installed aeration capacities for *P. monodon* and for shorter crops typical for *P. vannamei*.

Applying the guidelines (Table 2) to a 90-day crop of *P. vannamei* and comparing the results with those for the entire installed aeration capacity of 24 hp/ha operated for an average of 18 hr/day shows the benefit of staging aerator use. Aerator use would decline from 38,880 hp-hr in ponds aerated 18 hr/day throughout the crop to 28,140 hp-hr in ponds aerated by the guidelines—a reduction of 27.6 percent, assuming each horsepower hour delivered.

Table 2: Guidelines for Staging Daily Aerator Operation during Shrimp Grow Out

Modified from Mohanty (2001)			*Suggested for shorter crop*		
Days of Culture	*Aerators Operating (Percent)*	*Time of Daily Operation (hr/day)*	*Days of Culture*	*Aerators Operating (Percent)*	*Time of Daily Operation (hr/day)*
1-15	50	9	1-10	50	10
16-30	50	11	11-20	50	12
31-45	50	13	21-30	62.5	18
46-60	62.5	15	41-50	75	20
61-75	75	16	51-harvest	100	20
76-90	75	17			
91-110	100	18			
111-harvest	100	20			

Less aeration usually is needed in the daytime than at night. Generally, the period of lowest dissolved oxygen concentration will occur between 1:00 a.m. and 7:00 a.m. This suggests that the aerators to be operated daily during a particular period during the crop should all be operating during this critical 6-hr time interval.

Several recent reports have suggested new methods of aerating shrimp ponds such as wave (surge) aeration (Guan, Liu, Song, and Zhang, 2012), microbubble aeration (Liu, 2017), on-site oxygen generation and pure oxygen contact systems with aeration control by dissolved oxygen sensors (Nguyen and Matsuhashi, 2019; Tien, Matsuhashi, and Chan, 2019), solar-powered aerators (Chaithanakulwat, 2019), and submersible impeller aerators (Marappan *et al.*, 2020). These systems have yet to be perfected for use in shrimp ponds. For the foreseeable future, the best way to reduce energy demand of shrimp pond aeration is to enhance aeration management and improve the design of aerators. Some actions to improve aerator use follow:

i. Do not install more aeration than necessary. A good rule is about 2 hp/t for floating, electric aerators and 3 hp/t for long-arm aerators.
ii. Reduce energy use by current aeration practices by matching motors or engines with loads imposed by aerators.
iii. Further reduce energy use by staging operation schedule with proportion of installed aeration capacity in use and daily hours of operation increasing from stocking to harvest with respect to the amount of shrimp biomass.
iv. Promote use of factory-fabricated paddlewheel aerators rather than relying on farm-made aerators.
v. Instigate a program to show paddlewheel aerator manufacturers the benefit of adoption of more efficient paddle and paddle unit design.

vi. Encourage adoption of the 10-hp catfish pond aerator used in the United States or smaller versions of this aerator. This aerator is not under a patent, and it could be manufactured in Asia.

Energy Efficient Aerators

The average shrimp production to installed aeration capacity in shrimp ponds ranged from 153 to 597 kg/hp with an average of 295 ± 132 kg/hp. Hopkins *et al.* (1991) have clearly demonstrated that it is possible to obtain 500 kg/hp with factory-made, floating, electric aerators, and Paterson and Miller (2014) reported that such a ratio was typically achieved in Australia using such aerators. Two farms in Thailand using only farm-made long-arm aerators obtained y/hp ratios of 303 and 333 kg/hp. There was a rather wide range in energy use per tonne of shrimp. In some instances, values were between 12 and 15 GJ/t shrimp.

By staging daily aerator operation, it should be possible to lower energy use to 10–15 GJ/t even with aerators in current use in Asia. Adoption of factory-made aerators would decrease energy use at farms presently using farm-made, long-arm aerators, and particularly those powered by diesel engines. Finally, more energy efficient aerators could be made if manufacturers would use available information to improve aerator SAE. It seems reasonable to expect that a serious effort by the shrimp farming sector could reduce average energy use for aeration to 8 GJ/t or less.

Roughly 5 million tonnes of farmed shrimp are produced annually, and at least 70 percent of this production is from aerated ponds. A reduction of aeration energy use of 1 GJ/t shrimp (277.8 kW/hr/t) is equivalent to 972 million kW/hr of electricity on a global scale which would cost about USD 87.5 million. At an average farm-gate price of Rs.300-350/kg live weight of shrimp, the global farmed shrimp production has a value of around USD 21 billion. An energy savings of USD 87.5 million is 0.4 percent of the value of annual shrimp production. An energy reduction of 1 GJ/t also would lessen carbon dioxide emissions by ≈460,000 t CO_2 eq./year.

Chapter 12

Pumps in Aquaculture

Pumps are mechanical devices used to add energy to fluids so that mechanical (generally electric motors) is potential and/or kinetic energy of the fluid. Pump helps the water to be lifted to a point of higher elevation through increase in potential energy where as the increase in velocity or flow rate of water through a pipeline increases the kinetic energy of the water. In aquaculture enterprise, pumps are used to lift water from a low to a higher point against a gravitational gradient and water will flow through addition of energy. In hydraulic systems pumps are used to create a high pressure which allows the water to do work. Correct pump selection is important in aquaculture enterprises as the expenditure in pumping is one of the major costs. Poor pump selection increase the operating costs and increase the risk of pump failure at critical times.

Benefits of Pumps in Aquaculture

- ☆ Water can be lifted from a lower level to higher gradient
- ☆ Water of a tank can be exchanged as and when required.
- ☆ Water fed to farm is free from pollutants and unwanted materials because of screening.
- ☆ Though initial cost is more, maintenance cost is low.
- ☆ Controlled water management system
- ☆ Water can be aerated when sprayed into the pond by the ponds.

Even though several types of pumps are available, the most suited types of pumps for intensive aquaculture systems are listed below:

i) Centrifugal pump; ii) Reciprocating pump; iii) Submerisble pump; iv) Air-lift pump; v) Mixed flow pump.

These are employed in the culture farms to lift and feed the water to the production unit such as nursery and grow-out ponds.

1) Centrifugal Pump

A pump which raises the water from a lower level to higher level by the action of centrifugal force is called as centrifugal pump. These pumps require centrifugal force. If a vessel containing liquid is rotated about a point, the centrifugal force created will cause the liquid to raise to a point.

Principle

In centrifugal pumps, water which enters the center of casing (impellor) will rotate at high speed by means of an impellor which imparts both radial and tangential velocity to water. In doing so, the centrifugal force is created and water is thrown to the periphery. The water leaving the impellor is slowed down by casing. A portion of its kinetic energy is converted to potential energy that is pressure.

Important Parts

a. **Suction pipe:** The main function of the suction pipe is to suck the water from the source and then to pass it on to the delivery pipe. The suction pipe plays an important role in successful and smooth running of pump. A great care should be taken to make it water tight. A perforated foot value is connected at its bottom to avoid the entry of big and foreign materials in to the suction pipe. Bends in the suction pipe are to be avoided to minimize the water losses.

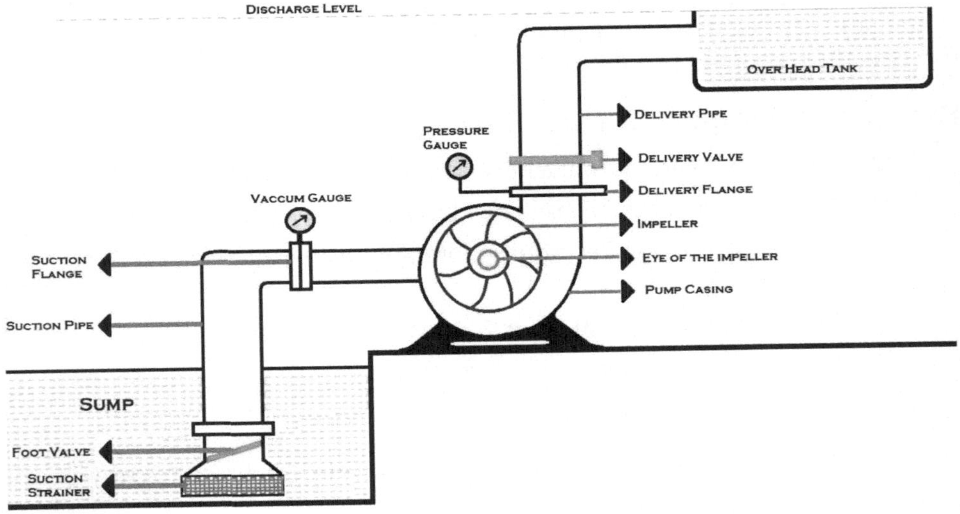

Figure 1: Parts of Suction Pipe

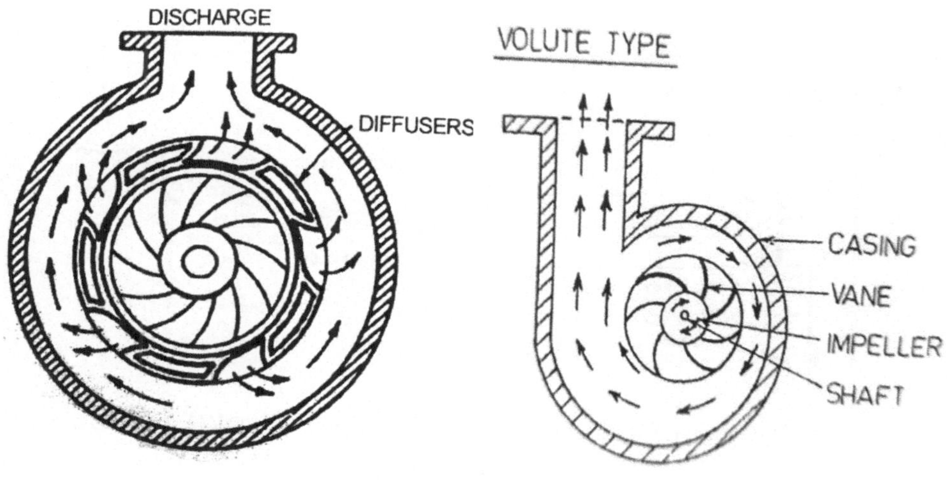

Figure 2: Diffuser Type Centrifugal Pump

b. **Motor:** It is very important part. Water from the suction pipe enters the motor (impellor). Impellor revolves at a designed speed depending upon the horse power of the pump. Due to rotation of an impeller surrounded by spiral casing, the kinetic energy will be created due to which water will be sucked and thrown to periphery; and thus, water enters in to the delivery pipe.

c. **Delivery pipe:** As the name itself indicates, the main function of delivery pipe is to discharge the water that comes from the motor to an outlet point. Sometimes, check valve is provided to protect the pump from water hammers. The size and length of delivery pipe depend upon the requirement and height between motor and water source.

2) Reciprocating Pump

A single acting reciprocating pump consists of a piston which moves inside the cylinder with the help of piston rod operated by fly wheel. A suction pipe and delivery pipe are connected to the motor. Piston moves inwards and outwards inside the cylinder in accordance with movement of fly wheel.

Working Principle

During the first stroke (*i.e.* suction stroke) the piston moves towards the right (from 0-180°) and thereby creating vacuum in the cylinder. This causes the suction valve to open; and thus, water enters the cylinder. During the second stoke (*i.e.* delivery stoke), the piston moves towards the left (180-360°); thus, increasing pressure inside the cylinder. Because of this pressure, outlet valve opens and the suction valve closes. During the process of opening and closing of outlet valve and its inlet valve, water that enters the cylinder will be forced into discharge pipe and thus, water is discharged.

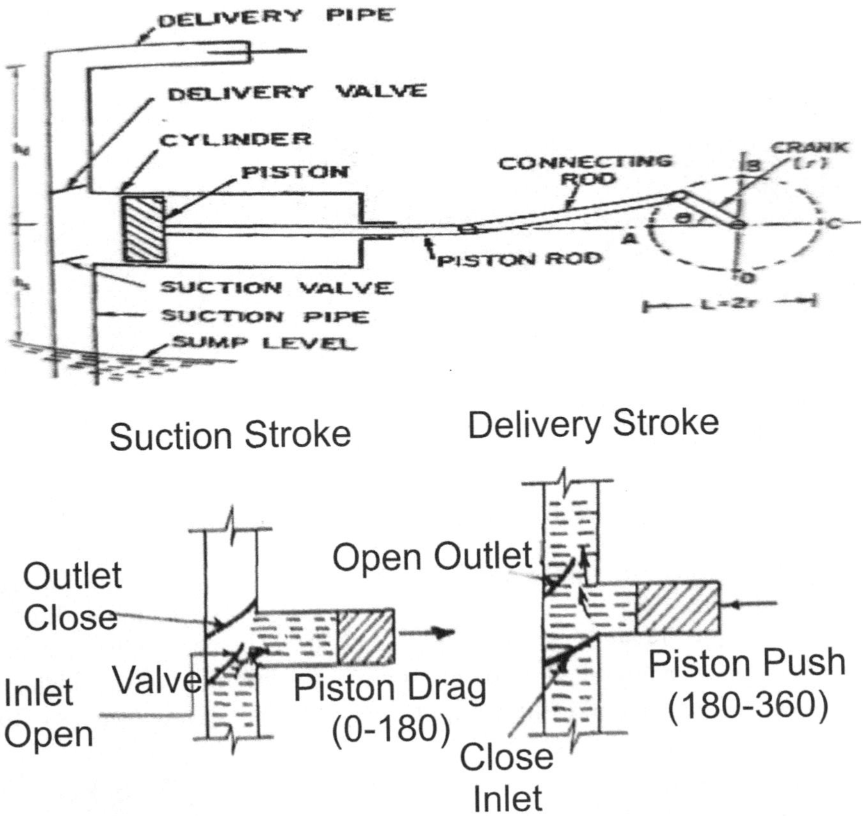

Figure 3: Suction and Delivery Strokes

This pump is also called as positive displacement pump as it discharges a definite amount of water during the displacement of piston. In double acting reciprocating pumps, there are 2 sets of suction and delivery valves.

$$\text{Discharge } (Q) = LAN/60$$

where,

L = Length of piston

A = Cross sectional area of piston

N = No. of revolutions per minute

Types of Reciprocating Pumps

- ☆ Simple hand operated reciprocating pump
- ☆ Power operated deep well reciprocating pump
- ☆ Single acting reciprocating pump
- ☆ Double acting reciprocating pump

3) Submersible Pump

A submersible pump is a pump which has a hermetically sealed motor close-coupled to the pump body. The whole assembly is submerged in the fluid to be pumped. The advantage of this type of pump is that it can provide a significant lifting force as it does not rely on external air pressure to lift the fluid.

4) Air-Lift Pumps

Water containing suspended matter and acids or alkali may damage other parts of pumps. The airlift pumps can be used successful in such situations. These types of pumps are suitable for lifting water from bore wells or deep wells. These are simple in operation and give least trouble because they have no remote/submerged moving parts. It consists of a vertical pipe or "educator pipe" placed in casing pipe. Air inlet pipe is placed in the educator pipe and its outlet upwards as shown in the figure. Top end of the airlift pipe is connected to the air compressor.

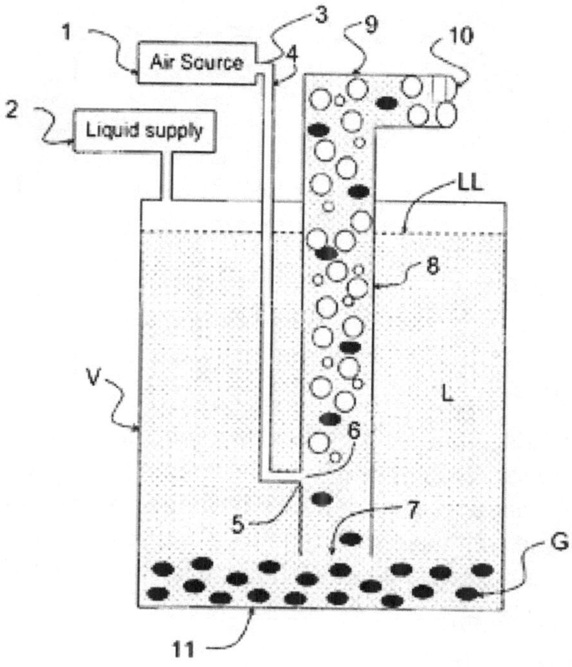

Figure 4: Air-Lift Pump

Working Principle

The compressed air is released through the air diffuser connected at the lower end of the air lift pipe in the educator pipe. This air is mixed with water and form bubbles having low specific gravity than water. Now, because of increased pressure, water starts moving in upward direction and water is finally discharged through the outlet. Only disadvantage of this pump is the cost of air compressor.

5) Mixed Flow Pumps

Mixed flow pumps are nothing but hybrid between the axial flow pumps and centrifugal pumps. The impellers are so designed as a compromise between a true centrifugal impeller and a true propeller type impeller. In this case, the energy is imparted partially by radial acceleration of the fluid and partially by the impeller mechanically lifting the water in an axial direction.

Due to this type of nature, the water can be lifted to a moderate height *i.e.* up to 7 meters and the discharge of water will be lesser to that of the axial flow pumps (*i.e.* 60 to 70 percent of axial flow pump water discharge).

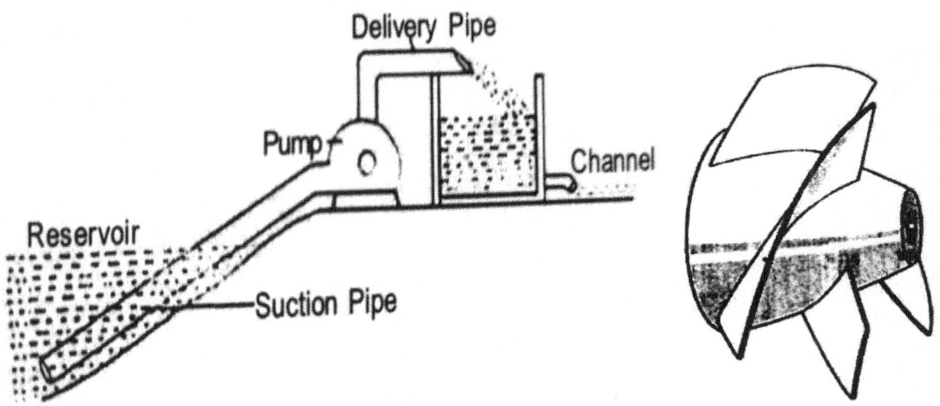

Figure 5: Mixed Flow Pump

Pump Selection

Correct pump selection for a particular application requires knowledge of the pump characteristics, the system requirements in terms of total head and discharge, and the characteristics of fluid which needs to be pumped Draw down characteristics of the well must also be known before a pump is selected for the well. Thus, before pump selection, the following factors need to be found out:

a) Total head required
b) Discharge required
c) Suction lift required

d) Characteristics of the liquid to be pumped
e) Whether the service is continuous or intermittent
f) Type of power source to be used
g) Space, weight and other limitations
h) Special requirements, if any.

For aquaculture purposes, generally low head high discharge pumps are considered most suitable. The requirement is met by the centrifugal pumps. However, in cases where the head requirement is less than 4m, the use of propeller pumps is considered more suitable, because the propeller pumps have relatively higher efficiency than the centrifugal pumps for the head up to 4m.

For certain purpose, the best selection of pump will be one that will operate at or near the peak efficiency. For this, the characteristics curves of the pumps be used. The characteristics curves relate the major pump parameters of head, discharge, efficiency and power requirements. Each pump has a unique set of these characteristics. For a particular situation, the required discharge and head should be estimated. Then superimposing these information on the characteristic curves of a pump, the efficiency at which it will run will be known. If the available efficiency is too far away from the peak value, then the characteristics curves of another pump should be tried.

After selecting the type of pumps, the capacity of pump is to be found out. This is required for the specific farming practice. This is done with the use of the rate of water exchange at the time of emergency. Generally, it can be calculated as 60 percent of the water holding capacity of the production system using the following formula.

$$\frac{\text{Water spread area} \times \text{Depth of water} \times 60}{12 \times 100} = \text{'X'}\ m^3/\text{hour}$$

Power Requirements and Efficiency

1. WHP (Water Horse Power efficiency) = (Q x H)/76

   Where, Q = Discharge (liter/sec)

   H = Total Head (m) $H = H_s + H_d + H_t$

   ☆ H_s = Suction head (lift)= Vertical distance between the water level and the center Line of the pump

   ☆ H_d = Discharge Head = Vertical distance between the center line of the pump and the point of free delivery of water.

   ☆ H_t = Head loss due to the resistance of the pipeline and fittings (including strainer, elbows, bends, reducing sockets, tees, valves, *etc.*).

2. SHP (Shaft Horse Power) = WHP/Pump efficiency
3. BHP (Brake Horse power) = WHP/(Pump efficiency X Drive efficiency)
4. IHP (Input Horse Power) = WHP/($n_p n_d n_m$)

   Where;

   N_p Pump efficiency

   N_d = Drive efficiency

   N_m = Motor efficiency
5. Kilowatt input to motor = IHP X 0.746

Chapter 13

Raceway System for Aquaculture

Raceways are culture units in which water flows continuously making a single pass through the unit before being discharged. These systems are also referred to as flow through systems. The residence time of the water in a raceway is very short, usually in the order of a few minutes instead of hours or days as in ponds. In the United States, raceways have traditionally been used for salmonid culture, but other species (*i.e.* catfish, oysters and clams) are also cultured in raceways to a lesser degree raceway culture has been traditionally practiced for hatchery production of fish for decades, and there are several good reasons for this. Environmental parameters (*i.e.* water quality, temperature, *etc.*) and water quantities are easier to manage in raceways than in pond systems. Flowing water flushes wastes from the culture units. Flowing water also forces the fish to exercise. Studies have shown that exercised fish have better survival rates when stocked into the wild (Wheaton 1977). The shallow water in raceways allows visual observation of the fish so that diet and/or disease problems can be promptly corrected. Finally, feeding and harvesting are generally easier in raceway systems. Feeding and disease treatment are more easily managed in raceway systems than in open systems or ponds. On the negative side, fish are normally cultured in very high densities in raceways, leading to increased risk of disease due to the stress caused by confinement and crowding. Better management skills are required for raceway culture.

Site Selection and Layout

Site selection for earthen raceways must be done with care. Raceway culture is water-intensive; therefore, probably the most important consideration is the water supply. Most raceway culture is in mountainous regions where gravity flow conditions can be taken advantage to supply the needed water to the raceway systems. In the trout culture industries in the Northeast and Northwest United States, freshwater springs are the main water sources due to their relatively low and constant water temperatures. The water quantity and temperature required depends upon the species cultured and the size of the operation.

In actuality, incoming water usually requires some degree of pre-treatment before it can be used. Pre- treatment may consist of nothing more than sedimentation, or it may consist of a combination of a number of processes including aeration, heating/cooling, degassing, or filtration. Depending on the effluent water quality, discharge water may also require post-treatment before it can be released into the environment.

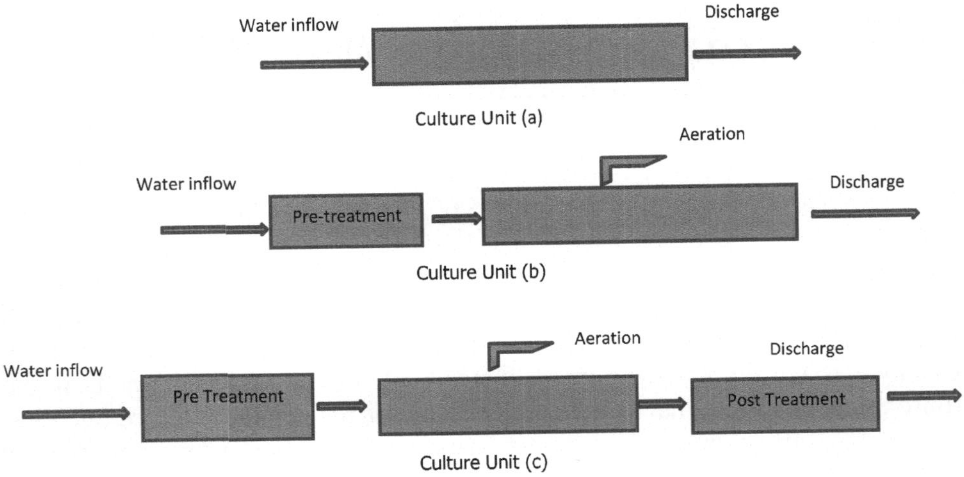

(a) No-O-sm treatment; (b) with pre-treatment and aeration; (c) with pre-treatment, post-treatment and aeration.

Figure 1: Single-Pass Raceway Systems

Multiple Raceway

Multiple raceway facilities are arranged either in series or in parallel. Series raceway systems can be used in regions where there is sufficient land slope so that the outlet for one raceway serves as the inlet for the next in the series. A 1-2 percent slope is considered adequate for good water flow (Pillay 1990). Waste buildup can be a serious problem in the series systems since the wastes from the upstream unit enter the next unit downstream, *etc.* Wastes tend to increase as the water traverses through the system. For this reason there is a practical limit to the length of raceways and the number of unit in series.

The waste problem is somewhat alleviated by arranging the raceway units in parallel. There are no culture units downstream receiving wastes from upstream units, therefore, waste buildup and aeration problems are minimized. However, the quantity of water required increases in direct proportion to the number of raceways. Thus, if pumping is required, operating costs are higher in the parallel arrangement.

Many large raceway farms use a combination of the series and parallel configurations.

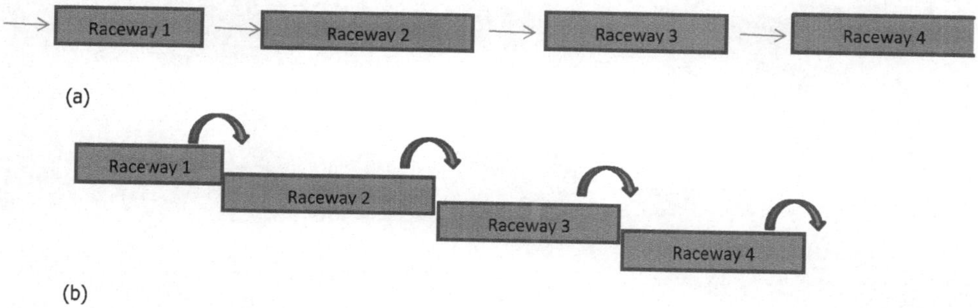

Raceway units in series (a) On flat ground (b) On sloping ground

Figure 2: Combination of Series and Parallel Raceway Units with Water Recirculation

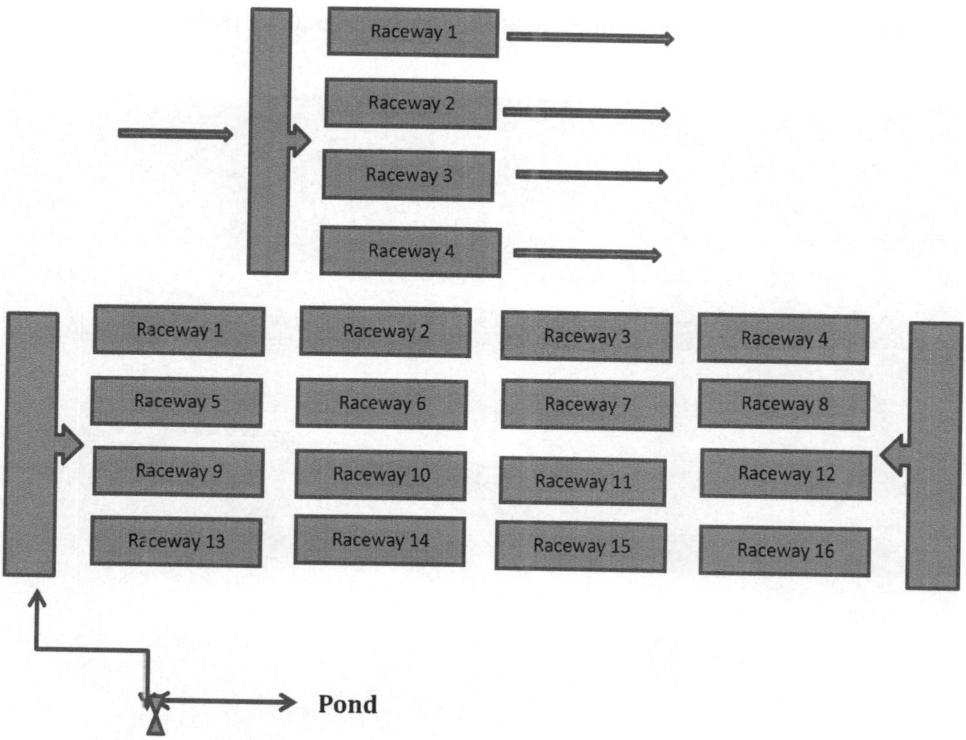

Figure 3: Combination of Series and Parallel Raceway Units with Water Recirculation

Figure 4: Raceways for Producing Fish with Recirculation (Felix, S., 2019)

Figure 5: Raceways for Producing Fish with Recirculation (Felix, S., 2019)

A commercial raceway system in Southern Arkansas for producing channel catfish was made. The facility consists of 108 concrete raceways, each 11m (36 ft) long by 5m (16ft) wide with a water depth of 1.2m (4ft). The raceway sections are arranged In three descending tiers, creating three oxygen-producing waterfalls. Supplemental oxygen is added at the inflow to each section with a liquid oxygen system. Water flows by gravity from a header tank, and water retention time in each section is about 6-7 minutes. The effluent from the raceway at the bottom of each tier discharges into pond and then circulates through a series of about 57 ha. of fish ponds to settle solids, oxidize nitrogenous wastes, and re-oxygenate the water. The water is eventually pumped back up into the header tanks at the rate of about 1.9 m/s (30,000 gpm).

In the Arkansas raceway facility, catfish weighing about 113g (0.251b) or more are stocked at 108 fish/m^3 (10 fish/ft^3) or about 15,000 fish in each raceway section. Large fish are stocked so that two crops of fish can be reared during a 200-day growing season. The facility produces about 907,000kg (907 tons) of catfish annually.

Construction Materials

Although earthen raceways are sometimes used, the majority are constructed from concrete or cement blocks, like the Arkansas raceway facility previously described. Earthen raceways are sometimes lines with waterproof liners to reduce water loss through leakage. Many small, experimental raceways are fabricated from concrete, wood, metal, fiberglass, plastic or other materials.

Water Supply

Raceways are designed to provide flow through for culturing very dense populations of fish. An abundant flow of good-quality water is essential to provide for the health of the animals and to flush wastes from the system. Water quality in a raceway is maintained by manipulation of the water flow rate and by adjusting other water treatment processes, aeration for example, in response to the demands of the cultured animals. Water quality tends to vary along the long axis of a raceway, and a distinct degradation of conditions is evident between the inlet and outlet. The specific flow rate required to meet the oxygen demands of the fish and flushing of metabolites is determined by the influent water temperature and dissolved oxygen concentration and by the oxygen consumption and ammonia excretion of the fish in the raceway.

Chapter 14

Filteration Systems in Aquaculture

Effective removal of suspended solids and organic matter from the water is called filtration. In aquaculture system inputs such as feed are provided for the aquatic organisms. These organisms excrete their faeces as output to the water. Thus, excess input and output (excreta) can become toxic to the aqua system causing reduction in growth of aquatic organisms. In the semi-intensive and intensive fish farming practices, this problem is prominent. These particle need to be removed (a) for the inlet water to the farm, (b) for the outlet water from the farm, and (c) for reuse of water in the farm. It is difficult to remove all the particles fully from the water. The colloidal matter and bacteria present in water cannot be removed. However, a certain proportion of particles is essential which depends on the design and function of filter. The inlet water is required to be filtered to remove higher concentration of particles that may reduce the growth of fish and may cause mortality. The outlet water is predominant with particle size < 30-40μm. The large number of small particles account for only a limited part of the total volume of discharged particles. Since the volume of particles is much more important than the number of particles on the recipient water body, it is of great importance to remove the few large particles. The density of faeces from fish farming varies from 1.005 to 1.2, which means that the faeces will settle in water. Hence, filtration is needed to reduce the effect of the outlet water on water conditions in the recipient body. For re-use system, filtration is mandatory to avoid accumulation of particles in the system and reduction in fish growth. The particles in the water can be broadly classified according to size as follows:

a. Soluble- <0.001 μm

b. Colloidal – 0.001 – 1 μm

c. Super colloidal- 1-100 µm

d. Settleable - >100 µm

Total solids (TS) in the water represents the total amount of particles in the water which may be expressed as total dry matter. But, total suspended solids (TSS) is the size of particles of > 0.45 µm. In aquaculture system, liquid-liquid and liquid-solid filtration system is important.

Filtration is accomplished by the different properties of the materials to be separated. The material properties include

(a) Density
(b) Particle size
(c) Electrical properties
(d) Chemical properties
(e) Magnetic properties

'Purification efficiency' is used to evaluate how efficiently a filter is functioning and to compare filter systems. This can be defined as follows.

$C_e = (C_{in} - C_{out}/C_{in}) \times 100$

Where, C = Efficiency (percent)

C_{in} = Concentarion of the actual substance entering into the filter

C_{out} = Concentarion of the actual substance coming out of the filter

There are several methods for removing particles from a water flow. Accordingly, filters are of different types as follows.

Filtration

↓

Mechanical filtration

A) Screen filtration
a) Stationary screen
b) Rotary screen
c) Vibration screen

B) Sand /depth filtration
a) Sand filter
b) Diatomacous earth Filter (DE filter)

Gravitational filtration / Settling

a) Sedimentation
b) Centrifugation
c) Hydroclone

Chemical filtration

a) Foam fractionation
b) Carbon adsorption
c) Ion exchange

Biological filtration

Biofilters

Mechanical Filtration

In aquaculture system mechanical filters are used mostly for the separation of liquids and solids and occasionally used for liquid-liquid separation. Principle is based on the differences in particle size of the mixture (or solution) components to separate one part from the other.

- ☆ Simple in operation
- ☆ Easy maintenance

Various filters are commercially available based on the size of particles.

1. Screen Filtration

a) Stationary Screen

- ☆ These screens are most widely used.
- ☆ It should be placed across the path of fluid flow.
- ☆ The screens having definite mesh size can stop particles larger than that mesh size. Gradually the screens will be blocked and head loss will increase until the screen is completely blocked with particles preventing any water flow through it.
- ☆ The mesh size of screens may vary upto 1.5-3.0 mm.
- ☆ When a screen is used to filter inlet water or outlet water in fish farming, a screen cloth of metal or plastic threads woven to the required mesh size is employed.
- ☆ The screen surface needs to be cleaned frequently to avoid blockage. Several methods make the screen self cleaning includes back flushing, vacuuming or mechanical vibration of the filter cloth. Out of these three back-flushing is very common.
- ☆ The screen material should be corrosion free like plastic or HDPE.
- ☆ It is important that the screen surface is smooth so that the particles are not crushed.
- ☆ It is in-expensive, easy to install and operate.
- ☆ It gives very good strength.

b) Rotary Screen

- ☆ It is used where screen clogging is a problem, where screen cleaning is expensive or, where it is not feasible to shutdown while the screen is cleaned
- ☆ Here, the screen is partially submerged in the water that is to be filtered. The Submerged section filters water passing through it.
- ☆ Continuous rotation causes a section of screen to alternatively filter and be above the water surface, where it passes in front of a back washing

system. This helps the screen to operate continuously and automatically with low head loss and low labor requirement.

- ☆ Rotary screens produce significant quantities of back wash containing a high concentration of solids.
- ☆ In general, rotary screens are more expensive than stationary screen but do not require much maintenance.

These are two basic types of rotary screens:

i) Axial flow

ii) Radial flow.

i) *Axial Flow Rotary Screens*

- ☆ Here flow rate is determined by the screen mesh size, head loss across the screen and cleaning efficiency.
- ☆ Head loss depends on percent of hole area in the screen, amount and characteristics of particulate material, efficiency of backwash, screen rotation speed and flow rate.
- ☆ Properly designed and operated screens should have a head loss in the range of 15-60 cm.
- ☆ Axial screens are relatively inexpensive, easily operated and do not require much labor for maintenance.

Disadvantage: These screens must be circular in shape.

ii) *Radial Flow Rotary Screens*

- ☆ Radial flow screens consist of a cylinder, the periphery of which is covered with a screen mesh.
- ☆ Fluid enters the screen axially at one end and exits radially through the mesh.
- ☆ Back washing unit is placed on the outside of the screen at the top.
- ☆ Radial flow screens normally operate with 30cm or less head loss.

2. Sand/Depth Filtration

a) Sand Filters

- ☆ Sand filters consist of layer/depth of sand or other particulate material through which water is forced to flow. This type of filtration is also called as depth filtration.
- ☆ During filtration, the particles too large to pass through the spaces between sand grains are trapped in the layer by several mechanisms such as:

i) The particles present in water may be too large to pass through the channels and holes created between the sand particles in the filter medium.

ii) The particles may settle both unaltered and due to flocculation and adhesion, and

iii) The particles may be adsorbed by chemical and physical forces that attract them to the filter mass.

- The first process will occur on the top layer of the filter medium where the particles get trapped. This will cause increase of head loss and filter becomes blocked more quickly. For the rest two processes to continue, a low water Velocity through the filter medium is highly essential.
- Effectiveness of the filter depends mainly on the grain size of the filter medium and flow rate through the medium.
- Based on the direction of water flow through the filter medium, the sand filters can be classified into:

  a) Up-flowing filter and

  b) Down-flowing filter
- The down flowing sand filter is most commonly used and also called as gravity sand filter. In this case, water containing particulate material enter at the top and passes down through the filter bed. Perforated pipe or floor provides a collection system for clean water. Gravity provides the necessary energy to induce the flow through filter.
- Flow rate through gravity filters is described as Darcy's equation.

$$v = K\, h/d$$

where,

v = Average velocity of flow through the sand filter(m/s)

K = Permeability constant ($m^{3/}s$ per m^2 of surface area)

h = Head (I.e pressure) (m)

d = Depth of the filter (m)

The continuity equation can then be used to find out the flow rate through the filter.

$$Q = A.V$$

where,

Q = Flow rate (m^3/s)

A = Top surface area of filter (m^3)

V = Average velocity of flow through filter (m/s)

- When a filter needs to maintain capacity and to prevent clogging of filter medium, back-flushing facilities must be provided. Backwashing is the process of reversing the direction of flow through a filter to remove trapped particulate materials. Effective backwashing requires fluidizing the bed where mechanical friction between particles removes materials attached to the filter medium producing better cleaning.

- Generally, amount of back-flushing water are in the range of 1-5 percent of the bulk-flow.
- In sand filters, instead of using one type of medium on the whole filter depth, *i.e.* a single size medium, media of several sizes (multimedia) may be used to increase utilization of the filter.
- In down flowing filter, the largest media are on the top and the smallest at the bottom. To avoid mixing of the media during back flushing, media with different relative density may be used. The largest media must have the lowest density and the smallest media have the highest density. Third bottom layer with even finer particles, such as crushed ilmenite or garnet, could also be included. Thus, each medium can be characterized by its grain size, uniformity of grain size, grain shape and relative density.
- Sand filter depths or 60 cm to 120 cm are common for aquatic system.
- Based on hydraulic capacity, the sand filters can be classified into a) slow sand filter, b) rapid sand filter and c) continuously back-flushed filter. Slow sand filters are unpressurized filters where as rapid sand filters are pressurized filters. In a pressurized filter, the medium may be put inside a sealed chamber with the same overflow arrangements as for an unpressurized filter.
- The preferred loading rate is 0.68 $L/s/m^2$ or less for slow sand filters, up to 1.4 $L/s/m^2$ for rapid sand filters and up to 5.4 $L/s/m^2$ for continuously back flushing sand filters.
- The slow sand filters and rapid sand filters have grains size of 2-5 mm and between 0.3 to 4 mm respectively.
- For 0.3 mm and 0.5 mm sand grain size, about 95 percent of particles down To 6μm and 15μm can be removed. A typical pressurized rapid sand filter has a capacity of 136 kg sand, A surface area of 0.29 m^2, a maximum pressure of 3.5 kg/cm^2 and a design flow rate of 238 L/min.
- Sand filters may also be used to lower the BOD (Biological oxygen demand) of culture system effluent. Sand filters also remove some bacteria from water.

b) Vacuum Sand Filter

Vacuum sand filters are similar to gravity sand filters, except increased head is achieved by pumping the water out of the under drains and creating a partial vacuum under the filter.

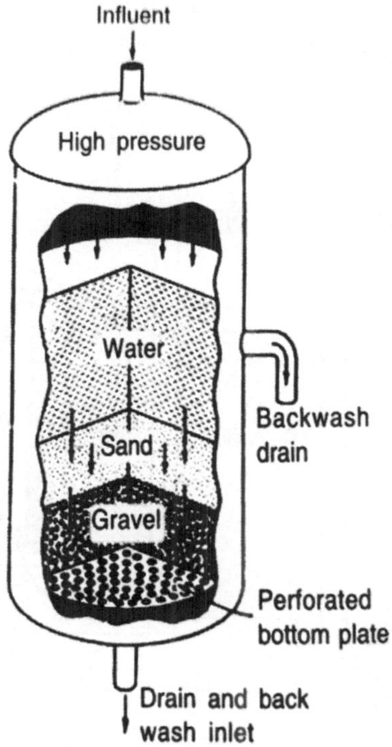

Figure 1: Vacuum Sand Filter

3. Cartridge Filter

- Depth filters can also be of cartridge type.
- The cartridge may be made of different materials *viz.* plastic, ceramic, spun,Fibre or resin bound fiber. This type of filter has definite depth. Wastewater is collected either inside or outside of the cartridge.
- This filter can be used for removal of particles below 1 μm. They are used on small water flows with low particle loads.
- Cost of operation is too high.

4. Diatomaceous Earth (DE) Filter

- These are mechanical filters used to get high water quality and low bacterial population.
- DE filters comprise a filter with a screen coated with DE (Fuller's Earth), a granular material of fossil diatoms to form a cake of DE which is put on the filter screen as main filtering medium. When water flows through the cake, the particles get trapped.

- It is capable of removing particles down to 0.1 μm.
- When the filter becomes dirty, it can be back flushed or replaced.

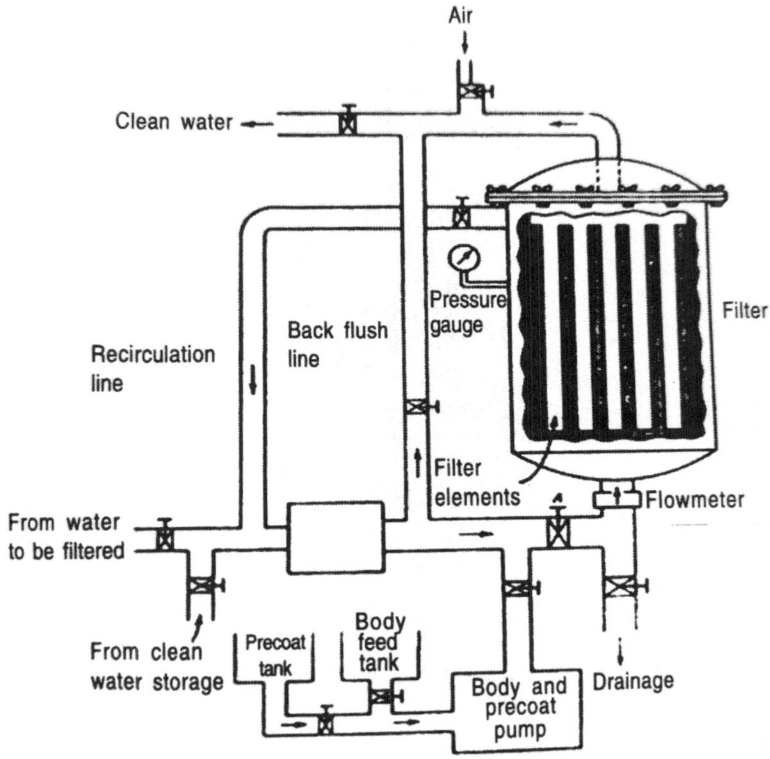

Figure 2: Diatomaceous Earth (DE) Filter

5. Gravitational Filtration

- The gravitational filtration utilizes the force of gravity to separate particles from water. The basic principle of such separation is the difference in relative densities of the particle and the water which cause the particles to travel downward. Generally, the relative density of particles varies from 1.005 to 1.2 where as that of water is 1.0. This causes the particles to settle. There are three gravitational separation techniques.
  i) Sedimentation
  ii) Centrifugation
  iii) Hydroclone

Each has possible application in aquaculture enterprises for separating solid from liquid.

i) Sedimentation

It is the process of allowing particulate materials having density greater than that of suspending liquid to settle out under gravitational force.

The sedimentation basins consist of 4 principal sections *viz.*

a) Inlet area
b) Settling area
c) Outlet area
d) Sludge area

- The inlet area serves to reduce turbulence and inlet velocity. The water then passes to settling area when the velocity must be very low. The clean water passes from the outlet section and the sludge settling area provides temporary storage for the material settled out. The flow pattern in such a basin used in fish farming is normally horizontal.
- The sinking velocity of the particle in sedimentation basin must be greater than the surface load or overflow rate for the settling basin.

  where,

$$V_s > Q/A$$

  V_s = Sinking velocity of particle (m/h)

  Q = Water flow through the basin (m^3/h)

  A = Surface area of the basin (m^2)

  Q/A = Surface load

In fish farm, the normal surface load is between 1m/h and 5 m/h.

Desirable depth for sedimentation basin is 1m and the width: length ratio is in the range of 1:4 to 1:8.

ii) Centrifugation

Centrifugation is used to increase the gravitational force experienced by particles during settling.

6. Chemical Filtration

- The chemical filters are primarily the adsorption units.
- Adsorption may be defined as a process of accumulation or concentration of substances at a surface or interface. The interface may be between two liquids, between a liquid or a gas, between a liquid and a solid, *etc.*
- Chemical or activated adsorption requires a chemical interaction to take place at the adsorption site between adsorbent and adsorbate (*i.e.* material being adsorbed).

- ✰ Charcoal, fly ash, aluminium oxide are the examples of different adsorbents. But active charcoal is found to be more efficient than others.
- ✰ Adsorption on active charcoal is a two or three step process depending on the location at which adsorption occurs.

First Step: Film Diffusion

- ✰ It is transport of solute through the surface film covering the active carbon.

Second Step: Pore Diffusion

- ✰ It is diffusion of the adsorbate within the pore of the adsorbent

Third Step: Adsorption

- ✰ It is the actual adsorption of the adsorbate on the surfaces of the activated carbon.

7. Ion Exchange

It is the electro-chemical exchange of ions that takes place between two materials, usually a solution and a solid that is insoluble in the solution.

Ion exchange is a surface adsorption phenomenon, which involves the interior structure of 3-dimensional molecules. Thus, it is both an adsorption and an absorption process and hence, is called as 'sorption'.

Ion exchange resins can be classified as:

i) Strongly acidic cations
ii) Weakly acidic cations
iii) Strongly basic anions
iv) Weakly basic anions

- ✰ The solution of correct resin for sorption of a specific compound depends on properties of compound.
- ✰ Ion exchange has possible application of purifying wastewaters from freshwater aquaculture systems.
- ✰ Removal of ammonium by ion exchange is very effective. The removal of some organic acids also appears to be feasible with ion exchange.
- ✰ Water softening is one of the common applications of ion exchange.
- ✰ Hardness of water, mainly due to presence of calcium and magnesium can be removed by passing the water through sodium based zeolite.

8. Foam Fractionation

- ✰ It is a process for separating or concentrating dissolved material by adsorbing one or more of the solutes at bubble surfaces. The adsorption concentrates solute at the bubble surface. Removal of the foam formed by the bubbles above the liquid surface removes the concentrated solute.

9. Biological Filtration

- Biological filtration is defined as the bacteriological conversion of organic nitrogenous compounds into nitrate. The several steps in this process are collectively referred to as the 'Nitrogen cycle'.
- The first step of conversion of nitrogen containing organic compounds into ammonia is completed before the material reaches the biological filter.
- Ammonification and de-amination occur in all biological filters influent is of pure inorganic source.

(NH_4^+) Ammonium (Highly toxic) —*Nitrosomonas* (bacteria)→ (NO_2^-) Nitrite (Less toxic) —*Nitrobacter* (bacteria)→ (NO_3^-) Nitrate (Non-toxic)

The primary purpose of a biofilter is conversion of ammonia to nitrite and then nitrite to nitrate.

- In the aquaculture system, ammonia is a highly toxic metabolic waste discharged directly by many cultured organisms and generated as a by-product by many bacteria. Nitrite is somewhat less toxic than ammonia and nitrate is considered nontoxic to most aquatic organisms. Even nitrate concentration up to 400 ppm does not have any toxic effect on aquatic organisms.
- The conversion of Ammonium to Nitrite is accomplished by autotrophic bacteria of genus *Nitrosomonas* in presence of O_2. Further oxidization of nitrite to nitrate is accomplished by *Nitrobacter* bacteria. These bacteria use energy, CO_2 and O_2 to produce organic compounds necessary for cell growth and metabolism. Biofilters consist of a porous solid phase on which nitrifying bacteria grow.
- The bacteria extract their nutrients, oxygen and other life needs from water passing over the solid phase.
- A simple biological filter consists of a box filled with rocks. Water enters the top, passes down through the filter and runs out through under drains in the bottom of the box.

Chapter 15

Recirculating Aquaculture System (RAS)

Aquaculture has been the fastest growing segment of Indian Aquaculture for more than 15 years and is projected to remain that way for the foreseeable future. This rapid growth rate is driven by several factors including:

i) Capture fisheries have reached their sustainable yield,
ii) Food safety concerns of consumers, and
iii) Consumer demand for high quality, safe aquatic products that are low in fat and high in protein,

Although the future of aquaculture is bright in India there also are risks. Regulation of water supplies (both quantity and quality), waste discharges and health regulations are becoming more and more onerous and costly to the industry. Competition for coastal sites, the public's concerns about environmental factors ranging from pollution to concerns about the visual "pollution" of aquaculture facilities on the shore, and the recreational use of waters that are also suitable for aquaculture are only a few of the clouds on the aquaculture horizon. These and other concerns are encouraging the aquaculture industry to move from open pond and cage culture systems to the more closely controlled recirculating systems. Typically recirculating (closed) aquatic production systems have higher capital costs than many of the extensive systems such as cage culture in natural waters and raceway and/or pond culture systems. However, when the control provided by recirculating systems and the benefits this environmental control provides in terms of fish health, waste capture, product quality, product availability, and other factors are considered—then recirculating systems become a more attractive business enterprise.

Though in its infancy and still considered to be a recent innovation, the basic technology of RAS has existed for over 65 years, with the first, pioneering RAS research activity being conducted in Japan in the 1950s.

Table 1: Timeline of RAS Development

Period	*Development*
1950s	Research on RAS first conducted in Japan
1970s	Roots of modern RAS through German program that demonstrated intensive carp production
Mid 1970s	Idea for RAS as commercial fish production was first fostered in 1970s Denmark
1980	First commercial RAS of European eel production was built in Denmark
Early 1980s	RAS design and technology used for catfish and eel production in the Netherlands
1980s	Marine RAS begun in China
1980s-1990s	Development of RAS in other European countries
2000-2020	Further development of RAS in Australia, Europe,North America and Other Countries

A significant acceleration in the development of RAS technology has been observed over the last two decades (Table 1), and RAS have become popular in recent years. RAS have been developed to grow fish where inadequate biophysical conditions, water scarcity, poor water quality, and unfavorable environment exist. RAS provide an alternative production method when environmental regulations, disease, land availability, salinity, temperature, and water supply prevent more cost-effective alternatives.

However, other factors stimulated the development and implementation of RAS. For example, RAS are increasingly being used for Mediterranean marine fish and salmonid production cycle, particularly for juvenile stages, before being transferred into outdoor grow-out systems, such as cages or flow-through raceways. In fact, RAS can be used for broodstock and seedstock production, which can support cage and net-pen aquaculture. In Europe and North America, RAS were developed as an alternative to cage culture of salmon.RAS have also been developed to culture exotic fish species, to avoid adverse effects on native species and biodiversity.

RAS can be categorized into five types:

(a) Hatchery and grow-out,

(b) Breeding,

(c) Long-term holding,

(d) Short-term holding, and

(e) Display.

Moreover, RAS can be incorporated into an "integrated agriculture-aquaculture" system, which is known as 'aquaponics'.

According to available scientific literature, the stocking densities of RAS range from 70 to 120 kg/m^3 with feed conversion ratio (FCR) values from 0.8 to 1.1. RAS can be of various sizes including small, medium, and large, with a large-scale RAS typically being able to produce 400-500 tons of fish per annum.

Challenges of RAS

- ☆ Despite opportunities for greater fish production through RAS, the current contribution of RAS to global aquaculture production is minor. In fact, RAS have not yet been widely practiced due to a series of social, economic, and technological challenges. Although aquaculture is highly concentrated in South and Southeast Asia (89 percent of global production), the adoption of RAS is rather slow in these regions, due to significant technological and economic constraints.
- ☆ Additionally, RAS are rather new features of the aquaculture sector in many countries, and thus, social acceptability is a still prerequisite for its wider adoption. The bulk of global aquaculture is broadly operated by small and medium farms, all enterprises which typically have limited financial resources.
- ☆ Key challenges for RAS are inadequate system designs and complex engineering approaches. Thus, RAS are commonly operated in highly developed countries (Australia, Canada, Europe, and the USA), rather than developing countries. Sophisticated equipment (*e.g.*, water temperature regulators, filtration, and sterilization units), advanced monitoring, sensors, and automated control systems are used in RAS.
- ☆ According to Murray *et al.* (2014), RAS are a costly engineering approach, with high initial investment to install and operation. Profitability determining factors of RAS are productivity, price of products, initial investment, and operating costs. Simple designs, whilst likely limiting overall productivity and stocking density, may help reducing capital and operational costs of RAS.
- ☆ High-energy use in RAS is one of key drawbacks, which increases operational costs. Typical energy consumption in RAS is 15-30 kWh/kg fish produced.

Need for RAS

(i) Freshwater scarcity is a facing challenge in many parts of the world, and competition for freshwater use between aquaculture and other users, such as agriculture, is rapidly increasing. Thus, food production is challenged to increase through efficient water utilization. Globally, aquaculture used 201 km^3 of freshwater in 2010, which is expected to reach 469 km^3 in 2050 (Mungkung *et al.*, 2014). Though aquaculture conceptually is not a water consumptive practice, as effluent waters can be reused for other purposes, it has a negative water footprint, primarily via water losses through infiltration and evaporation. The average water losses in freshwater aquaculture through infiltration is 6.9 m^3/kg. and a pond with seepage and evaporation loses 35 m^3/kg of fish produced.

(ii) On the other hand, RAS have the highest degree of water recirculation, which enable up to 90-99 percent of the water to be recycled (Badiola *et al.*, 2012; Dalsgaard *et al.*, 2013), and well-designed RAS are not associated with any major water losses. Water dependence in RAS is 93 percent lower than flow-through system (d'Orbcastel *et al.*, 2009), and 90-99 percent lower than conventional aquaculture systems (Ebeling and Timmons, 2012), which may help conserve water resources (Table 2).

(iii) Ultimately, water use in RAS is considerably lower than other aquaculture systems. Water use in RAS can be as low as 0.016 m^3/kg of marine fish produced (Tal *et al.*, 2009), and in general, water usages in RAS is less than 0.1 m^3/kg of fish produced (Martins *et al.*, 2010). Whereas it can be as high as 45 m^3/kg in extensive aquaculture (Verdegem *et al.*, 2006), and around 30 m^3/kg in a traditional flow-through trout farm (Bregnballe, 2015). In fact, water productivity in RAS (0.712 kg/m^3) is considerably higher than in all other aquaculture systems, including extensive aquaculture (0.01 kg/m^3) (Lemoalle, 2008; Verdegem *et al.*, 2006).

Table 2: Comparison of Freshwater use between RAS and other Aquaculture Systems

Aquaculture System	*Type of Farming*	*Water Use (m^3 water/kg if fish produced)*
Recirculating aquaculture systems (RAS)	Marine fish farming	0.016
	Fish Farming	0.3-3
	Next generation innovation	<0.1
	African catfish farming	0.5
	Eel farming	0.7
	Turbot farming	1.4
Pond Aquaculture	Intensive farming	2.7
	Pellet-fed use	11.5
	Extensive farming	45
Flow-through system	Trout farming	30

(iv) Waste from RAS is in the form of highly concentrated sludge that can be relatively easily managed and can be used as fertilizer for vegetable cropping (van Rijn, 2013), or at least anyway separated from farm effluents to prevent them entering into the catchment ecosystems.

(v) Accordingly, well managed RAS are not associated with water pollution and eutrophication, due to filtration, extraction, and concentration of excess nutrients. Eutrophication possibility in RAS is 26-38 percent lower than flow-through system (d'Orbcastel *et al.*, 2009), and 40 percent lower than net-pen system (Ayer and Tyedmers, 2009).

(vi) In addition to nutrients and catabolic waste products, aquaculture usually contributes to water pollution via the widespread application of antibiotics, chemicals, and fertilizers (Hall *et al.*, 2011). Globally, the average water pollution from aquaculture was estimated in 2010 to equate to 273 kg of nitrogen and 76 kg of phosphorus per ton of edible protein (Waite *et al.*, 2014).

(vii) Net-pen and cage culture discharge nutrients directly into the environment, which in marine aquaculture, have been estimated to increase up to six-fold by 2050 (Bouwman *et al.*, 2013).

(viii) Marine eutrophication from seafood production is likely to increase from 1.4 million tons in 2010 to 3.2 million tons N equivalent in 2050 (Table 3). Moreover, freshwater eutrophication from inland fish production will increase from 0.38 million tons in 2010 to 0.89 million tons P equivalent in 2050 (Waite *et al.*, 2014). Accordingly, it is not surprise that RAS have been indicated as one of the solutions to eutrophication and water pollution, while reducing environmental degradation (Zhang *et al.*, 2011).

(ix) RAS may be able to reduce, on average, 89 percent, 69 percent, and 100 percent of total ammonia nitrogen, nitrite, and nitrate, respectively by using modern technologies (Yogev and Gross, 2019). According to Chang *et al.* (2019), an effective design of integrated multi-trophic RAS could minimize aquaculture discharge, reduce marine pollution, and disease control.

(x) Globally, aquaculture occupied 18.8 million ha of land (12.8 million ha inland and 6 million ha coastal) in 2010 that is projected to increase to 44 million ha in 2050 (Waite *et al.*, 2014). Where there is a limited access to land and water, RAS are highly appropriate (Martins *et al.*, 2010), as RAS use less than 1 percent of land area than conventional aquaculture systems (Ebeling and Timmons, 2012).

(xi) RAS are not environmentally harmful in terms of habitat loss, due to requiring very little land to be used.

(xii) Exotic fish cultivation in RAS is possible with minimal ecological effects, In fact, fish escape is extremely unlikely in RAS due to the utilization of a closed system operation with proper protection, and thus, RAS provide very limited risks of biological pollution

Table 3: Comparison of Carbon Emissions from different Aquaculture Systems

Aquaculture System	*Country*	*Species*	*Emission (Kg CO_2) -eq/kg live weight)*
RAS (hydroelectric electricity)	USA	Salmon	3.73
RAS (typical electricity)	USA	Salmon	7.01
RAS (100 percent non-renewable energy)	Spain	Cod	16-27
RAS (50 percent renewable and 50 percent non-renewable energy)	Spain	Cod	12-18
Ras (electricity and oil fired heater)	Canada	Salmon	28
Net-pen aquaculture (power generator)	Canada	Salmon	2
Pond aquaculture (without land use change)	Bangladesh	Nile Tilapia	1.58
Pond aquaculture (with land use change)	Bangladesh	Nile Tilapia	1.81
Pond aquaculture (without land use change)	India	Major Carps	1.84
Pond aquaculture (with land use change)	India	Major Carps	2.12
Pond aquaculture (without land use change)	Vietnam	Striped Catfish	1.37
Pond aquaculture (with land use change)	Vietnam	Striped Catfish	1.61

(xiii) Although aquaculture is often dependent on wild-caught fry (*e.g.*, cod, eel, grouper, milkfish, shrimp, and tuna) with high levels of bycatch (Lovatelli and Holthus, 2008; Naylor *et al.*, 2000), RAS may contribute to reduce this occurrence, due to the possibility of developing hatchery operations for seed production.

(xiv) RAS are extremely efficient in terms of feed management and application, with minimal feed loss and a superior control of water movement and growing conditions.

(xv) RAS have their own biosecurity levels, but typically these include solid removal, biological and chemical filtration, water sterilization, disease prevention, and disinfection

(xvi) RAS are not associated with any major land use change like converting paddy fields or mangroves to fish farms, as their footprint is proportionally very limited and can be implemented in a variety of areas, including unused civil/industrial areas.

(xvii) RAS located in close proximity to markets can significantly reduce CO_2 emissions associated not only with natural habitat destruction, but also with fish transportation

Conclusion

Technological innovations should aim at developing simple design, low-cost RAS, which may help to establish small-scale, next generation RAS for responsible aquaculture practices, and to increase fish production and climate change adaptation. The potential for addressing food security in developing countries through RAS seems supportable. Besides high-value fish, adding different species, including carp, catfish, clam, milkfish, mussel, tilapia, and seaweed could provide diverse seafood production which may help more sustainable RAS operation Integrating tropical species (e.g., milkfish, sea cucumber, and sea purslane) with a simple and low cost RAS has the potential to increase seafood production and income of coastal communities in developing countries (Senff et al., 2020).Coastal communities, particularly in low-lying aquaculture producing countries, which are more vulnerable to climate change, should embrace RAS for adaptation to climate change. Community-based organizations, likely with the support of governmental and non-governmental organizations, may help capacity development and implement community-based RAS. Community-based RAS requires further technological innovation to facilitate accessibility via the minimization of initial capital investment and operational costs. Strong collaborations among these organizations with key stakeholders should aim at providing training, technical assistance, and empowerment of small and medium farmers as well as coastal communities for accomplishment of community-based RAS.

Multiple Choice Questions

PAPER 1

1. **Fresh water is defined when the salt content in it is less than ______.**

   (a) 0.5 ppt (b) 10 ppt
   (c) 20 ppt (d) 5 ppt

2. **Brackish water is defined when the salt content in it is in the range of______.**

   (a) 0.50-30 ppt (b) 35- 40 ppt
   (c) 0 ppt (d) None of the above

3. **Marine water is defined when the salt content in it is more than ______.**

   (a) 0 ppt **(b) 30-35 ppt**
   (c) 0.5ppt (d) None of the above

4. **______ includes breeding the aquatic organisms and raising the seed.**

   (a) Nursery ponds (b) Grow-out pond
   (c) Complete farming (d) None of the above

5. **Generally ______ types of ponds are used for seed raising.**

   (a) 1 **(b) 3**
   (c) 9 (d) None of the above

6. **______ is the culture management includes artificial feed along with natural food and fertiliser management.**
   (a) Semi-intensive culture (b) Intensive
   (c) Extensive (d) None of the above

7. **______is the culture management for maximum production from minimum quantity of water.**
   (a) Semi-intensive culture **(b) Intensive culture**
   (c) Extensive (d) None of the above

8. **______ involves one pond operation for either seed raising or raising to table size fish.**
   (a) Semi-intensive culture **(b) Restricted farming**
   (c) Extensive (d) None of the above

9. **______is the culture of single species.**
   (a) Monoculture (b) Restricted farming
   (c) Extensive (d) None of the above

10. **______culture involves 2 or more species culture together.**
    (a) Monoculture **(b) Polyculture**
    (c) Extensive (d) None of the above

11. **In polyculture, the species of ______ feeding habits are selected and stocked.**
    (a) Carnivorous **(b) Different**
    (c) Similar (d) None of the above

12. **An example for Composite fish culture is ______.**
    (a) Indian Major Carp (IMC) (b) Catfish
    (c) Murrel (d) None of the above

13. **______ is the culture of one sex either male or female of one species.**
    (a) Polyculture (b) Monoculture
    (c) Monosex culture (d) None of the above

14. **An example for Mono-sex culture is ______.**
    (a) Tilapia (b) IMC
    (c) Both (a) and (b) (d) None of the above

15. **______is the culture of fish along with other animals and/or crops.**
    (a) Composite fish culture **(b) Integrated farming**
    (c) Both (a) and (b) (d) None of the above

16. Examples of Integrated farming is/are______.

(a) Rice-fish farming,
(b) Fish- poultry farming,
(c) Fish-duck farming.
(d) All of the above

17. ______ is the culture practices in the water bodies of tropical climate.

(a) Cold water culture
(b) Warm water culture
(c) None of the above
(d) Cold weather culture

18. ______is the culture practices in the water body of higher altitudes.

(a) Cold water culture
(b) Warm water culture
(c) Both (a) and (b)
(d) None of the above

19. Total quantity of water required for a pond on an average per year is calculated using the following formula______.

(a) (Vp +V rf +Le + Ls + Lc+ Vra) m^3

(b) QTw = (Vp +V rf +Le + Ls + Lc- Vra) m^3

(c) Le + Ls + Lc

(d) V rf +Le + Ls + Lc+ Vra

20. Some examples for Surface water resources______.

(a) Streams
(b) Rivers
(c) Reservoirs
(d) All of the above

21. Ground water resources are ______.

(a) Wells,
(b) springs
(c) Both (a) and (b)
(d) None of the above

22. Miscellaneous resources of water for aquaculture ______.

(a) Rain water
(b) Run-off water
(c) Cycled water
(d) All of the above

23. Ponds which rely on rainfall to fill the need for water is called ______

(a) Sky Pond
(b) Earth Pond
(c) Cloud Pond
(d) All of the above

24. Water along with ______water is most suitable for aquaculture.

(a) Lake
(b) Well
(c) Pond
(d) None of the above

25. **Three major considerations in constructing fish ponds in non-irrigated and hilly areas are:**
    (a) Availability of large catchment areas
    (b) Availability of impermeable (non-porous) soil strata below water source and pond bed, and
    (c) Occurrence of sufficient underground water at a low depth
    (d) All of the above

26. **Surface run-off is maximum when type of catchment is______.**
    (a) Rock base or hilly terrain
    (b) Thickly wooded contour
    (c) Both (a) and (b)
    (d) None of the above

27. **The surface run-off is minimum when catchment is______.**
    (a) Sandy soil
    (b) Pasteur land.
    (c) Clayey soil with moderate percent of sand
    (d) All of the above

28. **Surface run-off depends on:**
    (a) Type of catchment and
    (b) Area of catchment
    (c) Both (a) and (b)
    (d) None of the above

29. **The volume of surface run-off can be calculated as following: Q = A x I × R, here Q refers to______.**
    (a) Q= Discharge of surface flow
    (b) A= Area of catchment, m^3
    (c) I= Intensity of rainfall, feet/h
    (d) R - No of Runs

30. **Cold water fishes can grow well in an optimum temperature range of ______**
    (a) 10 to 12°C
    (b) 25 to 35°C
    (c) 20 to 37°C
    (d) None

31. **Indian major carps thrive well at temperatures between______.**
    (a) 30 to 37°C
    (b) 10 to 12°C
    (c) 20 to 37°C
    (d) 10 to 37°C

32. **Adverse effects of turbidity are______.**
    (a) Decrease in light penetration.
    (b) Oxygen deficiency has been reported as a result of sudden increase in turbidity.
    (c) Gill choking occurs to fish seeds.
    (d) All of the above

33. Transparency of pond water can be measured by means of "secchi disc' and the values between______ are good for fish ponds.

(a) 25 to 70 cm
(b) 50 to 90 cm
(c) 20 to 30 cm
(d) All of the above

34. Acidic water (pH 5.0 to 5.5) can be ______to eggs, fry and adults.

(a) Healthy
(b) Harmful
(c) Neutral
(d) All of the above

35. The most common method of correcting the low pH (acidity) is by ______of the lime.

(a) Cation action
(b) Buffering action
(c) Iron replacement
(d) None of the above

36. Liming' Generally is applied in following forms:

(a) Lime stone ($CaCO_3$)
(b) Quick lime (CaO)
(c) Slaked lime or caustic lime or hydrated lime [$Ca(OH)_2$]
(d) All of the above

37. Liming of the pond serves the following purposes in the pond______.

(a) To correct and maintain pH of water (buffering action)
(b) To act as disinfectant
(c) To fix humic acid and inorganic (HSO) acid
(d) All of the above

38. In highly productive waters, the alkalinity ought to be over ______

(a) 200 ppm
(b) 100 ppm
(c) 300 ppm
(d) 500 ppm

39. In aquaculture systems, the alkalinity should be between ______in fresh water.

(a) 80 to 200 ppm
(b) 100 to 150 ppm
(c) 300 to 350 ppm
(d) 50 to 150 ppm

40. Alkalinity above ______adversely affects the spawning and hatching of carps.

(a) 200 ppm
(b) 100 ppm
(c) 300 ppm
(d) 500 ppm

41. High levels of free carbon dioxide (>16 ppm) leads to ______spawning.

(a) High
(b) Low
(c) Poor
(d) Better

42. **At Dissolved oxygen content of 0.3 to 0.5 ppm, fishes die. Oxygen depletion can be controlled by following methods like______.**
   (a) Exchange of water
   (b) Artificial aeration
   (c) Spraying of water on the surface of ponds or agitation of surface water
   (d) All of the above

43. **A permeable stratum capable to yield ground water under gravity is known as ______**
   (a) Air pump
   (b) Aquifer
   (c) Springs
   (d) Reservoir

44. **The silt particles get deposited in the ______ will be reduced.**
   (a) Canal
   (b) Water velocity
   (c) Both (a) and (b)
   (d) None of the above

45. **That velocity of water is called______**
   (a) Scoring velocity
   (b) Strange velocity
   (c) Silting velocity
   (d) None of the above

46. **That water velocity in the earthen canal is called______**
   (a) Scoring velocity
   (b) Strange velocity
   (c) Silting velocity
   (d) None of the above

47. **Hardness is the total soluble ______ salts present in water and expressed as its $CaCO_3$, equivalent.**
   (a) Calcium
   (b) Magnesium
   (c) Both (a) and (b)
   (d) None of the above

48. **Pond waters having hardness of ______ or______are satisfactory for growth of fish and do not require addition of lime.**
   (a) 13 ppm or above
   (b) 15 ppm or above
   (c) 15 ppm or below
   (d) 16 ppm or below

49. **Available phosphorus (P,O,) content in soil should be above______ for higher fish production.**
   (a) 50 ppm
   (b) 20 ppm
   (c) 60 ppm
   (d) 40 ppm

50. **Available nitrogen content in soil between ______ preferred for better fish production.**
   (a) 50 to 75 mg/100g
   (b) 5 to 75 mg/100g
   (c) 50 to 55 mg/100g
   (d) None of the above

51. **The ordinary gravity wells of______ m diameter.**
    (a) 3-7 m
    (b) 2-5 m
    (c) 1-2 m
    (d) None of the above
52. **which are constructed to tap water from the topmost water bearing strata are known as______ wells.**
    (a) Artesian
    (b) Non-artesian
    (c) Both (a) and (b)
    (d) None of the above
53. **The Manning formula is also known as______**
    (a) Gauckler–Manning formula
    (b) Gauckler–Manning–Strickler
    (c) Both (a) and (b)
    (d) None of the above
54. **______soil is required for fish pond and dyke construction.**
    (a) Sand
    (b) Loam
    (c) Cohesive
    (d) Non-cohesive
55. **______gases are important to the productivity of the soil and in turn of the fish pond.**
    (a) Ne,Ar,Cr
    (b) N,K,Al
    (c) N,P,Al
    (d) Ca,P,Al
56. **Mixed soils containing ______ percent clay is most desirable for fish pond construction.**
    (a) 20 to 30 percent
    (b) 10 to 20 percent
    (c) 40 to 50 percent
    (d) 5 to 10 percent
57. **______ found in the soil and water convert ammonium ion to nitrates by nitrification process.**
    (a) Pseudomonas
    (b) Nitrosomonas and nitrobacter
    (c) Aeromonas
    (d) Bacillus
58. **Available phosphorus (P,O,) content in soil should be ______ for higher fish production.**
    (a) Below 60 ppm
    (b) Above 20 ppm
    (c) Below 10 ppm
    (d) Above 60 ppm
59. **The ideal pH range of soil is ______.**
    (a) 6 to 7
    (b) 4 to 5
    (c) 7.0 to 8.5
    (d) 1.0 to 3.5

60. ______refers to the changes in the surface elevation of natural ground whether the ground is flat, sloping, undulating or hilly.

(a) Topography (b) Depth of excavation

(c) Elevation of water table (d) None of the above

61. ______of wind has direct influence while selecting a suitable site for fish pond construction.

(a) Distance **(b) Velocity and direction**

(c) Perimeter (d) Area

62. The most desirable wind speed in the selected site should be ______

(a) 4.5 km/h (b) 2.5 km/h

(c) 6.5 km/h (d) 7.5 km/h

63. ______of 10 to 15 are adjudged to be ideal for good soil fertility.

(a) C/P ratio (b) N/P ratio

(c) C/C ratio **(d) C/N ratio**

64. Pond bottom has a suitable site should have a coefficient of permeability(K).The value of K is

(a) 4×10 m/sec **(b) 5×10 m/sec**

(c) 6×10 m/sec (d) 7×10 m/sec

65. Soil texture determined by______method.

(a) Sieve (b) Bouyoucos hydrometer

(c) International pippete **(d) All of the above**

66. The ground slope of topography should not______.

(a) More than 2 (b) Less than 1

(c) Less than 2 (d) More than 1

67. ______can determine the kinds of pond which can be built based on whether the site is of low land or upland.

(a) Elevation (b) Depth of excavation

(c) Topography (d) Both (a) and (b)

68. Optimum temperature of______is found to be most suitable for fish to grow normally.

(a) 30°C **(b) 27°C**

(c) 15°C (d) 25°C

PAPER 2

1. **The art of determining the relative heights of points on the surface of the earth is called ______.**

   (a) Levelling
   (b) Surveying
   (c) Spirit levelling
   (d) None of the above

2. **Levelling deals with measurements of ______ plane.**

   (a) Horizontal
   (b) Vertical
   (c) Radial
   (d) None of the above

3. **______ is an imaginary line on the ground joining points of equal elevations.**

   (a) Level line
   (b) Contour
   (c) Both (a) and (b)
   (d) None of the above

4. **The difference in elevation between two points is proportional to the difference in atmospheric pressure at these points are called as ______.**

   (a) Barometric
   (b) Trignometric
   (c) Spirit
   (d) All of the above

5. **Vertical distances with respect to horizontal line are used to determine the relative difference elevation between two adjacent points are called as ______.**

   (a) Barometric
   (b) Trignometric
   (c) Spirit
   (d) All of the above

6. **The elevation of points are computed from the vertical angles and horizontal distances measured in the field are called as ______.**

   (a) Barometric
   (b) Trignometric
   (c) Spirit
   (d) All of the above

7. **The levelling instrument used to provide a horizontal line of sight are called as ______.**

   (a) Level
   (b) Levelling staff
   (c) Arrows
   (d) None of the above

8. **______ is mostly used for levelling survey from the engineering point of view.**

   (a) Tilting level
   (b) Dumpy level
   (c) Reversible level
   (d) Wye level

9. **Contours to be plotted are actually trace out in the field by locating and making a number of points on each ______.**
   (a) Indirect method **(b) Direct method**
   (c) Both (a) and (b) (d) None of the above

10. **The method which is suitable for contouring of small areas ______.**
   (a) Indirect method **(b) direct method**
   (c) Both (a) and (b) (d) None of the above

11. **Which method of contouring is known as "contouring by spot levels" ______.**
   (a) Indirect method **(b) Direct method**
   (c) Both (a) and (b) (d) None of the above

12. **Which method is cheaper quicker and less tedious ______.**
   (a) Indirect method (b) direct method
   (c) Both (a) and (b) (d) None of the above

13. **______ is the side slope for loamy soil.**
   (a) 2:1 (b) 1:1
   (c) 1:2 **(d) 2.5: 1**

14. **______ is the Side slope for homogeneous clay.**
   (a) 2:1 (b) 1:1
   (c) 1:2 (d) 2.5: 1

15. **______ are engineering works created through the moving and processing of massive quantities of soil or unformed rock.**
   (a) Earthwork calculation (b) Area calculation
   (c) Work Calculation (d) All of the above

16. **Cost of construction = Rate of earthwork x Quantity of earthwork**
   (a) Rate of labour x Quantity of earthwork
   (b) Rate of earthwork x No. of labours
   (c) Rate of earthwork x Quantity of earthwork
   (d) Rate of earthwork x Area of construction

17. **The slope of the ground along a given line is known as the ______**
   (a) Longitudinal slope (b) Latitudinal slope
   (c) Horizontal Slope (d) Vertical Slope

18. Which survey is done for large fish farms______.

(a) Plane table survey
(b) Theodolite survey
(c) Compass survey
(d) Chain survey

19. MSL full form = ______.

(a) Mean sea level
(b) Mean survey law
(c) Mean survey level
(d) Major sea level

20. ______ – F.S = last R.L – First R.L.

(a) B.S
(b) G.S
(c) H.S
(d) A.S

21. Which survey is suitable for both the horizontal and vertical angles measuring?

(a) Theodolite survey
(b) Compass survey
(c) Plane table survey
(d) Chain survey

22. The instrument used for plane table are ______.

(a) Plane table with stand
(b) Spirit level
(c) Plumbing fork
(d) All of the above

23. The levelling staff are graduated into ______.

(a) Metre
(b) Centrimetre
(c) Miilimetre
(d) None of the above

24. The length of the telescope staff ______.

(a) 4 m
(b) 5 m
(c) 7 m
(d) 2 m

25. The instrument used for the measurement of both horizontal and vertical angle ______.

(a) Theodolite
(b) Levelling staff
(c) Protractor
(d) None of the above

26. Line joining the intersection of cross hairs to optical centre is called ______.

(a) Line of collimation
(b) Line elevation
(c) Line of sight
(d) None of the above

27. The average height of the sea for all stages of the tide ______.

(a) Mean sea level
(b) Elevation
(c) Sea level
(d) Above sea level

28. **Plane table surveying is the ______ construction of straight line and angles.**
    (a) Graphical (b) Mapping
    (c) Measuring (d) None of the above

29. **The principle of plane tabling is ______.**
    (a) Line joining the point (b) Line joining the line
    (c) Kine joining the angle (d) None of the above

30. **Datum surface is ______.**
    (a) Arbitrary surface (b) Geographical surface
    (c) Plane surface (d) Sea surface

31. **Point where instrument is set up for observation ______.**
    (a) Instrument station (b) Non instrument station
    (c) Plane station (d) None of the above

PAPER 3

1. **Volumetric composition of inorganic composition soil mineral matter is ______**
   (a) 45 percent (b) 5 percent
   (c) 25 percent (d) None of the above

2. **Based on relative proportion soil separates there are ______ textural classes identified.**
   (a) 15 **(b) 12**
   (c) 18 (d) None of the above

3. **______ has very large pore space.**
   (a) Silt (b) Clay
   (c) Sand (d) None of the above

4. **______ has very high water holding capacity.**
   (a) Silt **(b) Clay**
   (c) Sand (d) None of the above

5. **Good loamy soil contains ______ percent of silt.**
   (a) 30-50 percent (b) 50-80 percent
   (c) 0-30 percent (d) None of the above

6. **The moist ______ soils are sticky and when these are in dried form they become hard.**
   (a) Sand
   (b) Silt
   (c) Clay
   (d) None of the above

7. **Which among the following is not a characteristic of sandy soil.**
   (a) Water holding capacity is low
   (b) Intermediate sized particle
   (c) Do not possess plasticity
   (d) None of the above

8. **The type of soil and it's composition at a site has direct bearing on______**
   (a) Productivity of pond
   (b) Water quality
   (c) Soil quality
   (d) None of the above

9. **Which is refers to the relative proportion of various size groups of particles in a given soil**
   (a) Soil colour
   (b) Soil porosity
   (c) Soil Texture
   (d) None of the above

10. **Which one is have large sized particles and pore spaces**
    (a) Clay
    (b) Slit
    (c) Sand
    (d) Peat

11. **To conduct the test in field which type is used.**
    (a) Disturbed sample
    (b) Undisturbed Sample
    (c) Both (a) and (b)
    (d) None of the above

12. **Soil samples for aquaculture are normally taken to a depth______**
    (a) 1.5m
    (b) 2m
    (c) 3m
    (d) None of the above

13. **Which particles are smallest in size and posses fine pore space**
    (a) Sand
    (b) Clay
    (c) Slit
    (d) None of the above

14. **Which particles have high water holding capacity**
    (a) Clay
    (b) Sand
    (c) Slit
    (d) None of the above

15. **Which type of sampling requires more physical and chemical analysis**
    (a) Disturbed sample
    (b) Undisturbed sample
    (c) Both (a) and (b)
    (d) None of the above

16. Which shows the property of somewhat between sand and clay

(a) Sand **(b) Silt**

(c) Both (a) and (b) (d) None of the above

17. In which type of sample is the natural structure is preserved

(a) Undisturbed sample (b) Disturbed sample

(c) Both (a) and (b) (d) None of the above

18. ______ is defined by the number of pores present within the soil.

(a) Soil consistency (b) Soil density

(c) Soil porosity (d) None of the above

19. The porosity of soil is determined by ______

(a) Movement of air and water within the soil

(b) Movement of air and water without the soil

(c) Movement of minerals and water within the soil

(d) None of the above

20. Soil porosity is influenced by

(a) Soil texture (b) Soil structure

(c) Both (a) and (b) (d) None of the above

21. ______ refers to the ability of the soil to stick to itself or other objects and to resist deformation and rupture.

(a) Soil consistency (b) Soil density

(c) Soil porosity (d) None of the above

22. Soil color is determined primarily by the ______ composition of the soil.

(a) Inorganic **(b) Organic**

(c) Mineral (d) None of the above

23. Pond ecosystem is a ______

(a) Forest ecosystem (b) Grass ecosystem

(c) Natural ecosystem (d) None of the above

24. During the day time, the algae produce oxygen by

(a) Respiration **(b) Photosynthesis**

(c) Transpiration (d) None of the above

25. **Find odd one:**
    (a) Water removal is to be done by pumping
    (b) Mechanical aeration is required to maintain dissolved oxygen levels.
    (c) A submersible pump has unsealed motor close-coupled to the pump body.
    (d) Positive displacement pump as it discharges a definite amount of water during the displacement of piston

26. **Which type of pond shape commonly used ?**
    (a) Circular (b) Oval
    (c) Rectangular (d) None of the above

27. **Provisions to increase oxygen content in water are highly necessary.**
    (a) Non-drainable pond (b) Drainable pond
    (c) Spring water ponds (d) None of the above

28. **Ponds depending on rain water are called as ______**
    (a) Sky ponds (b) Spring water ponds
    (c) Barrage ponds (d) None of the above

29. **Ponds supplied by springs and by rain water are known as ______.**
    (a) Barrage pond (b) Drainable pond
    (c) Spring water ponds (d) None of the above

30. **By-pass channels are used for the evacuation of excess water These are also called as**
    (a) Barrage ponds (b) Spring water ponds
    (c) Rosary system or Dispersion ponds (d) None of the above

31. **Culture practice of non drainable pond is ______ type**
    (a) Extensive (b) Intensive
    (c) Both (a) and (b) (d) None of the above

32. **Barrage pond is also known as ______ ponds**
    (a) Drainable (b) Spring water
    (c) Dam (d) None of the above

33. **Linked ponds without autonomous and individual water supply require ______ by-pass channel**
    (a) Single (b) Double
    (c) Four (d) None of the above

34. ________ is preferable to the preceding method and permits the individual supply of each of the ponds

(a) Linked ponds without autonomous and individual water supply

(b) Linked ponds without autonomous water supply

(c) Both (a) and (b)

(d) None of the above

35. Linked ponds without autonomous water supply require______ by-pass channel.

(a) Single
(b) Double
(c) Four
(d) None of the above

36. type of ponds is suitable in valleys with gentle longitudinal slope and the disposition is ideal as long as flow is sufficient.

(a) Spring water ponds
(b) Parallel ponds
(c) Linked ponds
(d) None of the above

37. In fresh water tanks, which type of ponds are the most preferred

(a) Rectangular ponds
(b) Square ponds
(c) Circular ponds
(d) None of the above

38. The length and breadth ratio of rectangular pond should be

(a) 3:1
(b) 2:1
(c) 1:1
(d) None of the above

39. The size of Nursery pond Is

(a) 0.04 to 0.06
(b) 0.02 to 0.06
(c) 0.05 to 0.06
(d) None of the above

40. The size of Marketing pond is

(a) 0.01 to 0.02
(b) 0.02 to 0.05
(c) 0.2 to 0.4
(d) None of the above

41. Drainage ditches constitute ______ percent of total productive area.

(a) 10 to 15 percent
(b) 15 to 20 percent
(c) 20 to 25 percent
(d) None of the above

42. What are the disadvantage of larger ponds?

(a) High cost and harvesting difficult
(b) controlling disease difficult
(c) All of the above
(d) None of the above

43. Name the factors deciding the size of ponds in aqua form

(a) Man power
(b) Equipment
(c) Both (a) and (b)
(d) None of the above

44. What are the advantages of small ponds?

(a) Harvesting easy
(b) Availability of oxygen is more in water
(c) Both (a) and (b)
(d) None of the above

45. Square and round ponds are not preferred because

(a) High cost
(b) Harvesting difficult
(c) Wastage of land
(d) None of the above

46. Important steps for planning a aquafarm are

(a) Layout planning.
(b) Man power planning and Material planning
(c) Both (a) and (b)
(d) None of the above

47. The typical size of nursery pond range

(a) 0.5-1 ha
(b) 0.05-0.2 ha
(c) 0.2-0.5
(d) None of the above

48. The typical size of spawning pond

(a) 0.05-0.2 ha
(b) 0.01 – 0.02 ha
(c) 0.01-0.05 ha
(d) None of the above

49. What is the constitution of embankment and drainage of total productive area

(a) 30-40 percent
(b) 25-35 percent
(c) 20-25 percent
(d) None of the above

50. Area of the storage pond can be determined by

(a) Amount of cultured species
(b) Stocking density.
(c) Dyke slope
(d) None of the above

51. Which pond size is consider for best management practise

(a) 0.5-1 ha
(b) 1-2 ha
(c) 1.5-2 ha
(d) None of the above

52. ______ percent of the total water spread area is generally reserved for nursery ponds.

(a) 10-15
(b) 20-25
(c) 5-10
(d) None of the above

53. ______ponds are most economical of the geometrical figures from the point view of construction.

(a) Square (b) Rectangle

(c) Circle (d) None of the above

54. About ______ percent of the total productive area is provided for stocking ponds

(a) 70-75 **(b) 60-65**

(c) 20-30 (d) None of the above

55. ______tanks are commonly used for nursery

(a) Square **(b) Round**

(c) Rectangle (d) None of the above

56. For fish culture where the harvesting is done by nets, ______ ponds are preferred

(a) Rectangle (b) Square

(c) Round (d) None of the above

57. The first stage in the construction process involves general planning. For this purpose, a topographical map of scale ______is required.

(a) 1:2500 to 1:2000 **(b) 1:2000 to 1:2500**

(c) 1:25000 to 1:20000 (d) None of the above

58. The next stage involves detailed planning of each pond. For this purpose, a ______work plan of larger scale is required.

(a) Topographic (b) Geographic

(c) Angularity (d) None of the above

59. According to Pruginin and Ben-Ari (1959), ______of earth is required per hectare for construction of pond.

(a) 2500-4000 Square meter (b) 2000-2500 square metre

(c) 2500-3000 square metre (d) None of the above

60. These pegs form a network of coordinates in which each peg is marked by a ______.

(a) Paper and pen **(b) Letter and number**

(c) Xerox copy and number (d) None of the above

61. Generally, about ______of total pond depth is obtained by excavation.

(a) 1/3 to 1(1/4) (b) 1/3 to 1/4

(c) 1/4 to 1/3 (d) None of the above

62. **The heavier the equipment, the greater is the packing coefficient, usually 20-25 percent. This means that when 1 m' of earth is required for the packed embankment, ______should be actually provided.**

    (a) 1.3-1.4 square metre
    (b) 1.2-1.5 square metre
    (c) 1.2- 1.15 square metre
    (d) None of the above

63. **______are stretched between the top of pegs and posts to mark the complete profile of the dyke with its correct height, width and slope.**

    (a) Tying
    (b) Strings
    (c) Pegs
    (d) None of the above

64. **is used to show the positions of various farm elements.**

    (a) Wet white lime
    (b) Dry white lime
    (c) Hot white lime
    (d) None of the above

65. **Dry white lime powder is used to show the positions of various farm elements like dykes, channels and ponds which are to be constructed. It is called as______.**

    (a) Excavation
    (b) Construction
    (c) Staking
    (d) None of the above

66. **The ______soil should be stocked in one place and it should not be used for any construction.**

    (a) Top ground
    (b) Bottom ground
    (c) Mid bottom
    (d) None of the above

67. **Initially lines are drawn according to the layout followed by ______ and ______Of the stakes or posts.**

    (a) Pegging and fixing
    (b) Fixing and scaling
    (c) Pegging and Marking
    (d) None of the above

68. **______ is used to show the positions of various farm elements.**

    (a) Charcoal powder
    (b) White lime powder
    (c) Marking with sand
    (d) None of the above

69. **______ is the basic structure involved in designing the construction of aquafarms.**

    (a) Dyke
    (b) Embankment
    (c) Channels
    (d) None of the above

70. **Foundation material should have low permeability to avoid excessive ______.**

    (a) Seepage
    (b) Water loss
    (c) Drying
    (d) None of the above

71. ______ forms the boundary of the entire pond system
    (a) Periphery embankment (b) Internal embankment
    (c) Dyke (d) None of the above

72. The purpose of periphery dyke is to prevent ______.
    (a) Soil erosion (b) Seepage
    (c) Overtopping due to flood (d) None of the above

73. Which is the secondary dike______.
    (a) Periphery dyke **(b) Internal dyke**
    (c) Channels (d) None of the above

74. ______ is Provided as a safety factor to prevent overtopping of dike.
    (a) Periphery dike **(b) Free board**
    (c) Berm (d) None of the above

75. Shear of the Dike is also called as ______.
    (a) Hydraulic gradient (b) line of saturation
    (c) Line of overtopping **(d) All the above**

76. Free board for periphery or main dike is normally ______.
    (a) 0.6m (b) 0.3m
    (c) 0.5m (d) None of the above

77. ______ flat portion of the land between heel edge of the dyke and edge of the excavation of main pond.
    (a) Berm (b) Slope
    (c) Core (d) None of the above

78. Berm is essential for which type of pond.
    (a) Nursery pond (b) Small pond
    (c) Large pond (d) None of the above

79. For construction of dyke,the soil must be free of ______ constituents.
    (a) Inorganic **(b) Organic**
    (c) Fertilizers (d) None of the above

80. Which soil is more suitable for construction of dyke?
    (a) Sand (b) Silt and clay
    (c) Both (a) and (b) (d) None of the above

81. ______ inclusion of clay is more suitable for dyke construction along with sand and silt.

(a) 10-20 percent
(b) 40-50 percent
(c) 20-30 percent
(d) None of the above

82. The finished height of the dyke should be about ______ cm above water for small ponds and ______ cm above water for large ponds.

(a) 30 and 50
(b) 20 and 35
(c) 50 and 30
(d) None of the above

83. The desired shape of dyke is ______.

(a) Square
(b) Rectangle
(c) Trapezoidal
(d) None of the above

84. The soil erosion of the rear side of the dyke can be checked by ______

(a) Cement pitching
(b) Grass turfing
(c) Both (a) and (b)
(d) None of the above

85. Desirable slope of seepage line of clay loamy soil is ______

(a) 5:1
(b) 1:4
(c) 3:1
(d) None of the above

86. Pond dykes should be constructed not above the ______ of thickness in layers.

(a) 20 cm
(b) 40 cm
(c) 30 cm
(d) None of the above

87. ______ dykes were suitable for brackish water farm.

(a) Square
(b) Rectangle
(c) Compact
(d) None of the above

PAPER 4

1. Grooves are made up of?

(a) Flash boats
(b) Cutoff wall
(c) Gate floor
(d) None of the above

2. The Sluices designed for a peek in flow of ______ percentage water exchange in grow out pond

(a) 100 percent
(b) 75 percent
(c) 50 percent
(d) None of the above

3. **Type of pipe used for water supply in fish pond are?**
   (a) Copper pipe and wooden pipes
   (b) Plastic pipe and rcc pipe
   (c) Asbasots pipe and GI pipe
   (d) All of the above

4. **Which connects main feeder channel of the pond system to source of water**
   (a) Surplus weir
   (b) Main sluice gate
   (c) Both (a) and (b)
   (d) None of the above

5. **Expansion of RCC?**
   (a) Recycle capacity canal
   (b) Recirculating cement concrete
   (c) Reinforced cement concrete
   (d) None of the above

6. **Which contain high density polyethylene?**
   (a) Grooves
   (b) Screens
   (c) Both (a) and (b)
   (d) None of the above

7. **Diameter of outer outlet pipe of Rearing pond is?**
   (a) 10 cm
   (b) 20 cm
   (c) 25 cm
   (d) None of the above

8. **Which is a two pipe device for draining of water during daily exchange ?**
   (a) Nursery pond
   (b) Rearing pond
   (c) Grow out pond
   (d) None of the above

9. **Name the shutters serve the purpose of controlling the quantity of water flow from 12 percent to 50 percent of water exchange?**
   (a) Nylon net
   (b) Wooden shutter
   (c) Both (a) and (b)
   (d) None of the above

10. **What are the components of Main gate?**
    (a) Floor
    (b) Cutoff walls
    (c) Fash boat and screen
    (d) All of the above

11. **The flow of water through the body of soil is called as**
    (a) Erosion
    (b) Seepage
    (c) Clay core
    (d) None of the above

12. **For stable soil, the seepage loss is around**
    (a) 0.12 cm/day
    (b) 0.17 cm/day
    (c) 0.13 cm/day
    (d) None of the above

13. Methods used for Seepage control

(a) Clay core
(b) Sluice
(c) Monk
(d) None of the above

14. **______ are being used to fill up the gaps in soil**

(a) Sand
(b) Clay particles
(c) Red Soil
(d) None of the above

15. Sealing pond dyke and bottom with

(a) Polyethylene/Plastic/Rubber liner
(b) Metal/Sand/plastic
(c) Plastic/polyethylene/Cement
(d) None of the above

16. Most Common method of seepage control

(a) Clay core
(b) Brick lining
(c) Biological plaster
(d) None of the above

17. Gley or, Biological plaster method was developed in

(a) USA
(b) Japan
(c) USSR
(d) None of the above

18. Is the process by which water is converted into gaseous stage and returned to the atmosphere as vapour

(a) Condensation
(b) Evaporation
(c) Radiation
(d) None of the above

19. As the temperature increases, the rate of evaporation

(a) Decrease
(b) Increase
(c) Both (a) and (b)
(d) None of the above

20. Temperature and rate of evaporation are ______ to each other.

(a) Inversion
(b) Proportional
(c) Indirectly proportional
(d) None of the above

21. As the surface area increases, the rate of evaporation

(a) Increases
(b) Decreases
(c) Both (a) and (b)
(d) None of the above

22. The rate of evaporation decreases with an increase in

(a) Condensation
(b) Rainfall
(c) Humidity
(d) None of the above

23. Increase in wind speed results in increased

(a) Rainfall
(b) Wind
(c) Evaporation
(d) None of the above

24. ______Is responsible for pushing sea water shore ward through river mouths into estuaries, creeks

(a) Tide (b) Wind

(c) Rainfall (d) None of the above

25. The swelling is at its maximum during new moon and full moon periods occurring at______ days interval

(a) 13 (b) 14

(c) 15 (d) None of the above

26. three types of culture systems

(a) Intensive, Extensive (b) Semi intensive

(c) Both (a) and (b) (d) None of the above

27. Control of predators is difficult in

(a) Pump fed system **(b) Tide fed system**

(c) Open water system (d) None of the above

28. ______refers to the changes in the surface elevation of natural ground

(a) Chromatography **(b) Topography**

(c) Geography (d) None of the above

29. ______ is considered most important for site selection

(a) Terrain (b) Soil

(c) Water (d) None of the above

30. Land lying between highly interlacing creeks branching off from the lowermost zones of the estuary is called

(a) Swamps (b) Estuaries

(c) Ocean (d) None of the above

31. Brackish water Estuarine areas are broadly classified into ______Zones

(a) 4 (b) 2

(c) 5 (d) None of the above

32. Inlets from the sea and coastal areas are mainly lying on the

(a) Margin of Bays (b) Edge of Bays

(c) Middle of Bays (d) None of the above

33. Disused salt pans are most suitable for

(a) Shrimp culture **(b) Prawn culture**

(c) Mussel culture (d) None of the above

34. The salinity of salt pan areas is around

(a) 50 ppt

(b) 40 ppt

(c) 20 ppt

(d) None of the above

35. Suitable for infrastrucuture of pond

(a) Higher elavated ponds

(b) Lower elavated ponds

(c) Both (a) and (b)

(d) None of the above

36. One of the most important aspect of coastal aquaculture

(a) Water management

(b) soil management

(c) Both (a) and (b)

(d) None of the above

37. A preliminary survey of the water source includes

(a) Its flow patterns and directions

(b) Rise and fall in tides

(c) Silt carrying capacity, Rate of evaporation

(d) All of the above

38. Supply of tide water at ______ days interval would be sufficient to keep the brackish water ponds fit for aquaculture.

(a) 20 to 25 days

(b) 10 to 15 days

(c) 5 to 10 days

(d) None of the above

39. Soils are porous mixtures of

(a) Inorganic particles

(b) Decaying organic matter

(c) Air and water and a variety of living organisms

(d) All of the above

40. Acidification of soil lowers pH to

(a) 4 or less

(b) 5 or less

(c) 6 or less

(d) 7 or less

41. Areas with low tidal fluctuation of 1 m or less are unsuitable for______

(a) Pump fed system

(b) Tide-fed ponds

(c) Open water system

(d) None of the above

42. Sea water is the main source of water supply to______

(a) Brackish water ponds

(b) Freshwater ponds

(c) Both (a) and (b)

(d) None of the above

43. Soil classification is normally based on particle size and some soil properties such as

(a) Plasticity (b) Compressibility

(c) Both (a) and (b) (d) None of the above

44. ______ is a process of estimation of total quantity of water required for an aquafarm.

(a) Water budgeting (b) Soil budgeting

(c) Both (a) and (b) (d) None of the above

45. ______is a ratio of the drop head between channel sectors located at a distance.

(a) Hydraulic slope (b) Depth

(c) Both (a) and (b) (d) None of the above

46. The most common type of water conveyance system is the..

(a) Open channel (b) Feeder canal

(c) Both (a) and (b) (d) None of the above

47. Which canals used to supply water from the main water intake to the fish ponds?.

(a) Feeder canals (b) Open channel

(c) Both (a) and (b) (d) None of the above

48. Which channels carrying the water to ponds?

(a) Feeder canal **(b) Protection canals**

(c) Both (a) and (b) (d) None of the above

49. Wetted perimeter (P)=.

(a) P = B + 21 (b) P=B+11

(c) Both (a) and (b) (d) None of the above

50. ______ is the vertical distance between the top of the channel wall and maximum designed water level in the channel.

(a) Free board (b) Slope

(c) Both (a) and (b) (d) None of the above

51. The mouth of the supply inlet should be covered to prevent the entry of ______

(a) Trash fish or predatory (b) Dog

(c) Cat (d) None of the above

52. **The metal pipes need to be protected with a coating of.**
   (a) Anti corrosive paint
   (b) Corrosive paint
   (c) Both (a) and (b)
   (d) None of the above

53. **Area of cross section (A)**
   (a) $A = Bd + zd^2$
   (b) $A= Bd^2$
   (c) $A=zd^2$
   (d) None of the above

PAPER 5

1. **Turbine aerators are also called ______**
   (a) Spray aerators
   (b) Surface aerators
   (c) Propeller diffuser aerators
   (d) Gravity aerators

2. **______ consume more oxygen per unit weight than large size of same species.**
   (a) Small fish
   (b) Large fish
   (c) Both (a) and (b)
   (d) None of the above

3. **______ are also referred to as flow through systems.**
   (a) Recirculatory Aquaculture System (RAS)
   (b) Raceway system
   (c) Intensive system
   (d) All of the above

4. **Multiple raceway facilities are arranged either in ______ or in ______**
   (a) Series
   (b) Parallel
   (c) Both (a) and (b)
   (d) None of the above

5. **______ are devices used to increase the concentration of dissolved oxygen in the aquaculture system.**
   (a) Pump
   (b) Propeller
   (c) Aerators
   (d) None of the above

6. **These are the floating propellers placed below the water surface. When the propeller rotates, it creates great turbulence and thus increase dissolved oxygen concentration in the water.**
   (a) Surface aerators
   (b) Spray aerators
   (c) Floating paddle wheel aerators
   (d) Turbine aerators

7. **A pump which raises the water from a lower level to higher level by the action of centrifugal force is called as______**
   (a) Centrifugal pump
   (b) Propeller pump or Axial flow pump
   (c) Mixed flow pump
   (d) All of these

8. **A ______ pump consists of a piston which moves inside the cylinder with the help of piston rod operated by fly wheel.**
   (a) Simple hand operated reciprocating pump
   (b) Power operated deep well reciprocating pump
   (c) Single acting reciprocating pump
   (d) Double acting reciprocating pump

9. **______ is also called as positive displacement pump.**
   (a) Reciprocating pump
   (b) Submersible pump
   (c) Airlift pump
   (d) None of the above

10. **In double acting reciprocating pumps, there are ______ of suction and delivery valves.**
    (a) 1 set
    (b) 2 set
    (c) 3 set
    (d) 4 set

11. **______ are suitable for lifting water from bore wells or deep wells.**
    (a) Reciprocating pump
    (b) Submersible pump
    (c) Airlift pump
    (d) None of the above

12. **______ is a pump which has a hermetically sealed motor close-coupled to the pump body.**
    (a) Reciprocating pump
    (b) Submersible pump
    (c) Airlift pump
    (d) None of the above

13. **______ are hybrid between the axial flow pumps and centrifugal pumps.**
    (a) Reciprocating pump
    (b) Submersible pump
    (c) Airlift pump
    (d) Mixed flow pump

14. **For aquaculture purposes, generally low head high discharge pumps are considered most suitable. The requirement is met by the ______.**
    (a) Reciprocating pump
    (b) Submersible pump
    (c) Airlift pump
    (d) Centrifugal pump

15. The term ______ is used to describe the energy in pumping systems and is how manufacturers represent the performance of their pumps as a function of flow rate.

(a) Total head (b) Suction head
(c) Discharge head (d) None of the above

16.______ can be used in regions where there is sufficient land slope so that the outlet for one raceway serves as the inlet for the next in the series.

(a) Series raceway system (b) Parallel raceway system
(c) Both (a) and (b) (d) None of the above

17. ______ is to suck the water from the source and then to pass it on to the delivery pipe.

(a) Suction pipe (b) Motor
(c) Delivery pipe (d) None of the above

18. Impellor is otherwise called as ______

(a) Suction pipe **(b) Motor**
(c) Delivery pipe (d) None of the above

19. ______ are mechanical devices used to add energy to fluids.

(a) Pumps (b) Aerators
(c) Propeller (d) None of the above

20. ______ is the most important critical water parameter in aquaculture system affecting growth, feed efficiency, susceptible to disease and ultimately the fish production.

(a) Dissolved oxygen (b) Carbon dioxide
(c) Temperature (d) pH

21. ______ aerators inject air or oxygen into the water in the form of bubbles.

(a) Spray aerator **(b) Diffuser aerator**
(c) Surface aerator (d) Turbine aerator

22. Which type of aerator facilitates the volatization of undesirable gases ?

(a) Paddle wheel aerator (b) Turbine aerator
(c) Surface aerator (d) Diffuser aerator

23. Standard test for aerators are conducted in concrete basins of clean water at ______ temperature.

(a) 20°C (b) 30°C
(c) 42°C (d) 50°C

24. **A commercial raceway system in southern arkansas for producing ______ was made.**

(a) Tilapia (b) Common carp

(c) Mackerel **(d) Channel catfish**

25. **______ slope is considered adequate for good water flow.**

(a) 1-2 percent (b) 3-7 percent

(c) 6-7 percent (d) 9-10 percent

26. **______ has a advantage that it can provide a significant lifting force.**

(a) Airlifting pump **(b) Submersible pump**

(c) Mixed flow pump (d) Both (a) and (b)

27. **The moving part of centrifugal pump is called ______**

(a) Impeller (b) Volute

(d) Diffuser (d) Suction nozzle

28. **In a centrifugal pump the water enters the pump______**

(a) At the top (b) At the bottom

(c) At the centre (d) From sides

29. **______ are used in fine screen so that they do not clogged up.**

(a) Spray aerator (b) Gravel bed aerator

(c) Automatic strainer (d) Surface aerator

30. **Discharge of a centrifugal pump is ______**

(a) Directly proportional to N (b) Inversely proportional to N

(c) Directly proportional to N_2 (d) Inversely proportional to N_2

31. **Dissolved oxygen along with ______ plays important role in metabolism of fish.**

(a) Temperature (b) Hardness

(c) Salinity (d) Alkanity

32. **Rate of ______ Consumption varies by fish with water temperature and environmental DO**

(a) Oxygen (b) Feed

(c) Water (d) Fish

33. **By creating a large contact area between the water and the ______ a good gas water exchange will occur**

(a) Feed **(b) Air**

(c) Waste material (d) Fish

34. Gravity aerators are ______ and are most economic when a natural head exist.

(a) Expensive | **(b) Inexpensive**
(c) Costly | (d) Cheap

35. Which one is the floating propeller and placed below the water surface

(a) Surface aerator | **(b) Spray aerator**
(c) Wheel aerator | (d) Diffuser aerator

36. Which one of the propeller can cause physical damage to fish.

(a) Diffuser aerator | **(b) Turbine aerator**
(c) Paddle wheel aerator | (d) Spray aerator

37. Position of aerator in the pond should be decided based on______ direction.

(a) Water flow | **(b) Wind**
(c) South | (d) North

38. "B" factor of waste water is

(a) 0.65-0.75 | (b) 0.45-0.50
(c) 0.95-1.00 | (d) 1.00-1.05

39. Which aerator will reduces the daily fluctuations range of pH value

(a) Paddle wheel aerator | (b) Spray aerator
(c) Turbine aerator | (d) Diffuser aerator

40. Suction pipe is the important parts of the

(a) Centrifugal pumps | (b) Motor
(c) Reciprocating pumps | (d) Submersible pumps

41. In ______ aerator water is pumped through a nozzle at high velocity.

(a) Paddle wheel aerator | (b) Gravity aerator
(c) Surface aerator | (d) Spray aerator

42. In floating paddle wheel aerator the dO concentration of the water ______

(a) Increases | (b) Decreases
(c) Leap | (d) Lessioning

43. In hydraulic system pumps are used to create a ______

(a) Low pressure | **(b) High pressure**
(c) Moderate pressure | (d) None of the above

44. Reciprocating pumps operate by drawing ______ into the chamber.

(a) Liquid (b) Pressure

(c) Electricity (d) Heat

45. ______ pump is also called velocity pump.

(a) Reciprocating (b) Rotary displacement

(c) Centrifugal (d) Screw

46. ______ is used to describe the energy in pumping systems.

(a) Depth of water **(b) Total head**

(c) Section head (d) Discharge head

47. Vertical distance between the water level and the center Line of the pump______.

(a) Total head (b) Discharge head

(c) Section head (d) Head loss

48. Vertical distance between the center line of the pump and the point of free delivery of water.

(a) Discharge head (b) Total head

(c) Head loss (d) Section head

49. Water from the suction pipe enters the______.

(a) Motor (b) Suction pipe

(c) Delivery pipe (d) Entrance pipe

50. Function of delivery pipe is to discharge the water that comes from the motor to an outlet point is ______.

(a) Motor (b) Suction pipe

(c) Delivery pipe **(d) Outer pipe**

51. Aeration is the addition of ______ to water.

(a) O_2 (b) Co_2

(c) No_2 (d) NH_3

52. Too much oxygen to the water will also be toxic for the______.

(a) Plant (b) Bacteria

(c) Crustacean **(d) Fish**

53. ______ are devices used to increase the concentration of dissolved oxygen in the aquaculture system.

(a) Pipe (b) Fillter

(c) Aerator (d) Motor

Index

C

D

E

S

T

9 789359 192055

Printed at Repro India Ltd.